The Absolute

The Absolute

Granting this power to humanity, will give it
the harmony and balance to banish the
deviances exposed in everyday life.

No me gusta el estilo confuso como
escriban, pero las verdades que
dicen es lo que siempre han dicho
los "mysterious schools"

Carolina Fuentes
Catherine Fabbro
Mari Carmen Ortiz Monasterio

The Absolute
First English Edition: October 2012

Copyright © 2011, 2012 Carolina Fuentes, Catherine Fabbro,
Mari Carmen Ortiz Monasterio
Reg. No. 03-2011-1201 10575400-01

ISBN Print Version: 978-0-9860259-0-7
ePUB: 978-0-9860259-1-4
ePDF: 978-0-9860259-2-1
Library of Congress Control Number: 2012950385

Printed by:
Bookmasters, Inc.
30 Amberwood Parkway
Ashland, OH 44805
USA

Cover design: Lorena Arriola
Editorial assistance: Bogar Vallejo and Claudia Gamboa
Layout and editorial design: Editorial Página Seis, S.A. de C.V.
Morelos 1742, Col. Americana
44860 Guadalajara, Jalisco
México

Translated from the book *El Absoluto* in Spanish:
D.R. © 2011 Carolina Fuentes, Catherine Fabbro, Mari Carmen Ortiz Monasterio
Reg. No. (Spanish version) 03-2011-1201 10575400-01
ISBN 978-607-7768-36-4
Reg. No. (English version) 03-2012-101511335700-01

The authors recognize the importance and the responsibility they contract upon the delivery of this book and therefore invite all readers to send their questions. We emphasize that this is the only channel through which it is possible to have access to the studies of Konocimiento Kósmico (Kosmic Knowledge) and with the persons within them. We are not established in any area. All communication is by way of this medium:

libroelabsoluto@yahoo.com
or *www.promethe.mx*

Printed and made in the United States.

Dedication

With humility and respect, we thank the Greatest Entity that mankind has known, for the help that, through his teachings, we have received throughout our journey on the Earth, since the beginning of the HTimes until the present day. We are grateful for the Supreme Gift which he bestowed upon us: the Konocimiento (Knowledge) and the Kommunication with the Konceptos (Koncepts). We wrote this book with the intention of taking our first steps within the Absolute Konocimiento and with the longing that this awakens every human being to his true reasons.

He has seen us grow and develop and fail again throughout the centuries. Mankind has demonstrated arrogance, rejection, and aggression towards his teachings because of our lacks of humility and acceptance of our errors. And in spite of it all, he has always been there, ready to give us solutions time and again.

Without his aid, humanity could not have advanced, and much less be aware of and project itself towards reasons that are beyond our incipient comprehension.

We offer our joy and elation for him having granted us the appreciation of form and color, and for the very Konocimiento that was not accessible to man.

We wish to thank our Brothers who with their commitment, efforts, and dedication, have opened the way so that the Konocimiento (Knowledge) of this Third HTime remains on the face of the Earth. Without their help, this book would not have been possible. We offer them our most sincere thanks for helping us on this path.

And to our companions on the journey within the Konocimiento (Knowledge):

Xoroam Cortés and Peter Gagnon

To Marion, our Guiador (Guide), whom we hope to meet again someday.

*Our thanks to those who with their support,
made the realization of this publication possible.*

Ana Lucía de Teresa
Gerardo Ruíz Mateos
Elvira Velázquez
Frank Fabbro
Marc Blondeau
Carlota Gedovius

Preface

The studies which appear in italics were taken from the pages of the original notebooks. They are part of the studies and have been copied exactly as they were delivered, or as they have been compiled throughout the years in which the Konocimiento Kósmico (Kosmic Knowledge) has been developed.

The language which the reader will find in this book is the beginning of a new idiom. The words represent concepts, and within themselves have different levels of meaning which will amplify and reveal itself throughout the course of the HTiempos (HTimes).

Some words are written differently since that is how they were given, and there is a reason for this. Within this English translation, some words have been left in the original Spanish, as this sound has a vital reason to be. In these cases, translations have been provided, as marked within parentheses.

Words are forms and doors; they will open a different world in the direction of the Konocimiento (Knowledge). Many of the words that are found in this book have a meaning distinct from that which is currently used. There are other words which are new and will help open the mind to new reasons and teachings. For example, the term "Kommunication" signifies the primordial and vital connection we have with the Konceptos (Koncepts) and Supreme Entities that form and constitute us, and which activate all of the processes of development and evolution in humans and in the Universe.

The authors recognize the importance and the responsibility they contract upon the delivery of this book and therefore invite all readers to send their questions. It is emphasized that this is the only medium through which it is possible to have access to the studies of Konocimiento Kósmico and with the persons within them. The studies are not established in any area. All communication is by way of this medium:

libroelabsoluto@yahoo.com or www.promethe.mx

Prolog

The Ideas you will find in this book are the exposition of a journey toward a gift granted to humanity by the Maestro, and which is the beginning of the Konocimiento (Knowledge). This will give us the clarity necessary in order to comprehend our reasons.

The Konocimiento will give man the keys to reject the diversity of theories that comprise his environment, since many of them lack a complete truth, and a half-truth is a lie.

Many minds have sought without understanding the principle and beginning of the Supreme Reason, which is man's innate spirituality; for that reason they search tirelessly. That is why we say that the reason of the encounter with his Creator is what matters.

The part of the Konocimiento which is delivered to us is to open the konsciousness; but to receive it, we must only deserve it. It is enough to open what man calls mind-brain; it is thus that the wisdom that science proclaims is born. Sabiduría (Wisdom) has imposed a barrier which impedes his development and evolution, because this has been utilized in his nefarious attempts to take over the human mind; this wisdom also has been utilized to have slaves. Sabiduría (Wisdom) seeks the eternity of our current matter.

Humanity is full of arrogance for wanting to conquer the Universe; all of this is due to the confluence of Huestes (Hosts) that have believed themselves sovereign throughout all of the HTimes and thus have damaged evolution.

We can lose ourselves in thousands of philosophical theories but the exact point is: to have humility and submission so that these studies are delivered, which, being that they are so extensive, have been impossible to present in their entirety in a few pages.

They gave us the mandate to deliver them because the times are short to learn them. With this book we are giving the keys which will be doors in order to open and develop the studies until we have larger, more ample, and unique answers.

We hope to awaken in people the curiosity which is the path of the quest.

We are not a religion but we speak of its origin; religion is the return to the origin in order to understand the error that came to fruition on the basis of this.

We hope dear reader that the promise that we lay before you slowly opens your sense of investigation; if this is achieved, the objective of the Idea that is born is also accomplished; if the vision is widened, konsciousness is expanded.

If the Idea is not planted we will not know what to seek, but if the Idea is present, this will go on opening up the way and thus we will overcome the limits that the environment which we live in has imposed on us.

These pages humbly try to deliver what was delivered to us and we hope with longing that this Third HTime be the awakening of Konsciousness.

We know the responsibility of the same and the gift we were given when they were delivered to us, to be able to know them, understand them, learn them, assimilate them, and thus give them to humanity by means of this book which was made with humility and respect towards the Konocimiento.

Chapter 1
The Beginning

A man spent his life asking the cows where they came from.
And they all responded: "Mooo."
Tired, he asked the people where they came from.
And they answered: "Mooo."
Question: What is the difference between a man and a cow?
CONCLUSIONS

Helena hung up the phone and turned on the television to continue watching the overwhelming news about the earthquake in China. She touched the small mole between her eyebrows, as was her custom when she felt uneasy. She felt a deep sadness as she watched the events that had presented themselves in such a devastating way. The transformations that by law must occur in the Earth in this Third HTime made her think that time was getting shorter; it seemed to her that everything was accelerating in these moments. She saw that mankind, in its insensitivity and desire to transform the environment, was provoking damages to itself and still didn't grasp the full extent of its attitudes. Helena knew that worse than what was appreciable on the physical level was the damage being done on the Kosmic level.

She pondered how humanity had arrived at this delicate point. She thought of how the Universe has a carefully-sequenced development, and how each of the actions in progress in it has a purpose that must be completed within an allotted time. Helena imagined the Universe like an accordion that unfolds itself, giving entrance to new reasons that have evolution as their objective. All of the changes in the Universe (or Fusion or Creative Expansion), are subject to laws that manifest themselves in determined moments. Humanity, being part of the Fusion, is ruled by these same laws that project themselves within the Earth in a specific time. However, when the lessons that we need to learn are not completed in the agreed-upon period, we lag behind the course of the other events taking place in the Fusion. If the human race didn't reach the advances programmed for this Third HTime, another cataclysm could occur in order to prevent us from doing more damage and to shake us out of the fantasy in which we live: the continuum of human thought that has filled us with the illusion that we're the owners of all that surrounds us.

Helena remembered what she had learned in the studies: *'[...] I tell thee that when one cycle finishes and another is established, the majority of men show themselves to be clumsy and senseless but not so a small group of a few keen-witted men who move towards other moments in their existence. In the coming years, as thou sayest in the material, the solar formation or its structure will suffer profound changes that will not be comprehended by scientists. This will modify the structure in all of the orders of what thou callest Planicio Terráqueo* (the Earth). New species of plants and animals will appear, and a new structure will be consolidated within what thou callest microcosmos. Some seas will have a red tint in their waters, and this will surprise the pequeños (little ones)'* (13 Jan. 95). (*Earthly Plane. The Konocimiento explains us that the Earth is not round, but flat. It is due to our perception of the third dimension that we see it thusly. The multi-dimension in which we live presents it to us as round.)

Helena thought of how humanity currently is lagging behind the reasons specific to the Kosmo that we live in. She knew that there is an exact moment for everything, and in getting ahead of or falling behind the reasons we are meant to live, we provoke some of our own problems such as illnesses, disturbances in our konsciousness, and karma. She said to herself, 'to be in harmony with the reasons that govern this HTime, we have to open our mind, which is at this moment unaware of the reasons that govern our development. We have to adjust ourselves to the Kosmic reasons.* (*Reasons that rule our evolution.) If we don't, all that we live is just an imposition of our self-interests and a rejection of what we need to do.'

She reflected upon her responsibility. Right now humanity has a chance to once again know the truths of its existence. The Konocimiento (Knowledge) that she had been studying all these years needed to be delivered to all mankind; this was the opportunity that was being granted to the human race in this Third HTime. She contemplated the erroneous ideas that currently form our world: that the Earth and our matter* belong to us; that we only live once; that we have the right to explore all that exists within the Fusion, believing that we are the only sentient beings within it and that we are able to decide for ourselves. 'But,' she said to herself, 'we don't even know what we are, nor why we exist.' (*Our body is comprised of various layers, some composed of physical matter, such as organs, tissues, etcetera, and others of non-physical or subtle matter, such as the Ego, the Perespíritu, among others.)

She drank the last sip of her coffee and reflected on how different life had been for humans in the First HTime, when the Planicio Terráqueo (Earth) hadn't yet been affected by mankind's muddled thoughts and actions. She remembered a Konocimiento (Knowledge) that she had studied: *'man is not owner, he is a tourist for a period of time in this sphere. This, he has falsified: my earth, my homeland, etcetera. But he is unaware of it; he knows not where he comes from, nor where he is going to. [...] The vital reference points [...] are: from where and to where are we going.*

What did man do to himself? He separated HIS *truths from* THE TRUTH. *If once there were civilizations that achieved great heights of development and projected in Time, what happened? Could it be that certain groups corrupted and disrupted others* (physically and mentally) *in order to control them? (24 Sep. 88).* Who could they be?'

She pulled back her abundant chestnut hair, rose and put on her jacket. She looked in the mirror, noticed the mole on her forehead that contrasted with the paleness of her skin and stepped out into the garden. The plants were a great company for her and she felt content here in this place that she had planted and cultivated. When she had arrived the previous year to live in this little cottage in the hills outside of Remoulins, she noted that it needed various repairs, but had decided to first fix up the garden. She had designed it with care, following the colors and forms of the plants so they would have harmony and movement. The old dwelling that she had inherited from her grandfather was in the middle of a small olive grove that birds filled with their songs. In summer, the fragrances of lavender, pine and cedar wafted through the hills on the warm, dry breeze. The climate in the southeast of France allowed her to enjoy the flowers that she so enjoyed during most of the year. The town of Remoulins was situated on the other side of the river from the city of Avignon, where Helena had been born in the summer of 1968. Sometimes she couldn't believe how many changes she had experienced in her life. And how many more were yet to come?

While she walked around the space pulling the weeds, she said to herself, 'without a doubt the Earth is a place of great beauty and harmony, full of sounds, colors, forms and aromas that we don't appreciate. Unnoticed by us, great Kosmic Entities complete their labors in it, according to their responsibilities and level of evolution. Why do we work so hard to destroy our home? To damage the habitat that we have been given is to scorn the source that manifested us in it. Why this eagerness to alter and modify the Planicio Terráqueo (Earth)? What is it that they want? The leaders of the hidden hierarchy that controls mankind's thoughts advocate the development of the material world, telling us that it's for our benefit. And even more, now they seek to conquer the Universe, to extend the "glory" of humanity to all points of the horizon when we can't even take care of the home we have. This supposed well-being is fairly ironic, as we only need to look at the consequences of these recent events.'

After strolling around the small garden she felt more at ease. She returned to the house and went upstairs to her bedroom to bathe. When she finished, the air was still chilly so she dressed in a lavender-colored sweater and a pair of comfortable pants. She then took the ring that had been resting on the bedside table and placed it on her index finger.

Entering the kitchen Helena lit the old cast-iron stove to prepare another cup of coffee. In the mornings she enjoyed sitting at the table that her grandfather had carved. She had painted the walls of the kitchen in a delightful shade of green that cheered her. The house was old and a little worse for the wear; it still needed quite a bit of work, but Helena was very grateful to have it because here she found the calm she needed to be able to study. After the experience of living in the jungle and the meager conditions she had lived in for so many years, the house was a great gift.

She looked at the ring that she wore on her right hand; it was a little large for her finger. The light that entered through the window in this moment made the stone sparkle with multiple shades of green. Observing her long, thin hands, she wondered how different they had been before and if she had worn the ring on the same finger. She smiled at the idea. Although there were still some things that she needed to remember about that past life, this one didn't have the least bit of importance. The essential part she had understood, was that in this lifetime she had to deliver the truth that had been hidden and distorted so many times in the past.

She remembered the vision in which she had seen the ring for the first time. It had happened in the summer when she was nine years old and on vacation with her family at the beach, on the Costa Brava in Spain. They had travelled from their home in Avignon to spend a restful week there and everyone was overjoyed to be able to get out of the tiny house they lived in. A client of her father Bernard had invited them to spend the week with his family on the coast. Her father was a man dedicated to his job and the responsibilities of his family. He was curt and fixed in his ideas. He shared brief moments with Edouard, the eldest son, and with Phillipe, the second. Helena, the youngest of the three children, had been left in the care of her mother by her father, who felt no affinity with her and didn't understand her in the slightest. Her mother Jocelyne was a woman who had not had many aspirations in her youth; she had only awaited the moment in which she could marry and form her own family. For both her parents, Helena was a problem they didn't know how to solve. It wasn't easy, since she was very restive. They realized that she had no intention of adapting to life as they conceived it.

Helena was uneasy and always full of questions that seemed absurd to them. "If god exists, what is he like? What does he think? Do plants think? Where do we go when we dream?" These were some of the questions that Helena had posed to her parents. But they never knew how to respond and she had not found anyone who could explain things without always ending up repeating the same story: that we live, die and then go to heaven to have a good time. Helena insisted that, "There must be a reason why the Universe exists and why I'm alive, just like the animals and plants." As the years passed, and she was unable to find someone capable of answering her queries, her frustration grew until she decided that it would be better to just keep her concerns to herself.

As a young girl she had understood that she could do things that other people couldn't. Sometimes she knew what other people were thinking, who was calling before answering the phone, or anticipated if something was going to happen. She had sometimes even been able to move objects with her mind. Her parents became frightened at these events and they scolded her incessantly, telling her to control herself and to stop doing that. They had even threatened to take her to the psychologist to see if he would be able to make her fit in. Helena had felt isolated and rejected because of abilities that seemed natural to her, but provoked fear in others. She ended up spending her free time in her room, reading, drawing, or talking to her brother Phillipe, who was the only one that accepted her.

When she was nine, and the chance to go to the beach at Blanes in Spain had arisen, she was happy; she longed to get away from home. Her mother had planned everything. This was the first vacation the family had had in three years and she wanted it to be perfect. For Helena, the trip represented the possibility of seeing new things, of getting to know other places and other faces. She had passed the long hours in the car transfixed by the changes in the landscape, by the smells and colors distinct from those of Avignon. But when they arrived at their destination, her parents watched and controlled everything that she did and she had felt trapped yet again.

One day the entire family was at the beach. Helena, who was a very good swimmer, had felt the urge to go for a dip in the ocean. Anxious to get away from her parents who were so suffocating, she dove in, but soon found herself caught in a strong current that pulled her away from the shore. She had to swim all the way to the other side of the bay in order to return. She was exhausted when she finally got close to the beach again but didn't realize that a powerful wave was about to break right on top of her. The wave caught her and spun her around several times and just when she was finally able to reach the surface, another broke and Helena barely had time to take a breath before being pulled under again. In this moment she had remembered what her brother Phillipe had always told her, "Waves come in series of three." She wasn't sure if she would be able to survive another; she needed air. She tried to pull herself together and get to the surface. When she finally made it, she only had enough time to duck under again before another wave crashed on top of her. In these instants, her awareness had changed and she had entered into a different perception of time in which everything played out in slow motion. Her thoughts had ordered themselves and she realized that she wasn't afraid. She entered into a strange vision, not knowing where she was or who were the people she saw. After what seemed to be a very long time, everything turned to black. When she regained consciousness, she found herself on the beach, hearing a faraway voice calling her. She opened her eyes to see Edouard standing over her, taunting.

"Hey, sis, look what that little wave did to you. What happened to your powers?"

Phillipe had pushed him away, saying, "Back off, leave her alone," and moved closer to Helena. When she had recovered her breath, she grabbed him by the arm, asking him to sit next to her so that she could tell him what she had seen.

"When the third wave came, I dived back under; I barely took a breath. Suddenly everything went black and I felt my consciousness leave my body quickly; then I saw myself in a place I didn't recognize. It was in the desert; there was a big house with a patio in the center and a small pond surrounded by fig trees. I heard a lot of voices shouting and I turned to see a group of people, mostly men. They were dressed in long tunics; some of them were intricately embroidered while others were simpler. The men seemed to be arguing about a decision that needed to be made. There was a woman seated in the middle of the group who attracted my attention. She had pale skin and auburn hair, and was dressed in a long green gown that was embroidered around the neck and sleeves. She looked very worried. She was playing nervously with a ring that she had in her hands. It had a round green stone mounted in a thin gold band that was engraved with some kind of a geometric design; I don't remember exactly what it looked like. One of the men moved closer to her, threatening her. At that instant, she turned her face and our gazes met. You're not going to believe this Phillipe, but she had the same mole that I have on my forehead, and when I saw her eyes, I was so shocked to see that they were so much like mine that it woke me out of the vision. But I was able to hear her tell me something about the ring, or about finding someone. After that, everything was muddled and I only heard that fool Edouard making fun of me. Don't ask me why, but I think that the woman was me. Do you think that's possible, Phillipe?"

"I've heard that in some cultures they say that we live and die many times. Maybe that was what you saw, Helena," answered Phillipe.

Pensive, Helena asked him, "Who could she be and what did she want to tell me?"

They both sat in silence observing the immensity of the sea, trying to find an answer.

Phillipe was five years older than Helena. There had always been a strong connection between the two. It was as if they were both tuned to the same frequency where their thoughts and feelings flowed without interference and each was able to clearly understand the other. He had assumed the responsibility of caring for her. Helena demonstrated an intuition and psychic ability that scared other people, and they assailed her for it. At school it was common that Phillipe defended her from the classmates who were always taunting her, since for them Helena was just a dim-witted girl who didn't pay attention to her lessons.

During classes, Helena was bored and entertained herself imagining detailed scenes in other locales, scenes that seemed to be from a long-lost time. She generally snapped out of her reveries when someone pulled her hair or the teacher chided her for not paying attention. Once in a great while Helena reacted to the provocations of the other students, imagining a way to pay back the aggressions, as was the case with a classmate who had been yanking her long hair and bothering her for quite some time. She secretly wished that he would fall and break his arm, and then it happened just as she had imagined. Frightened, Helena ran to tell her mother, but she only received a scolding for telling lies and a visit to the parish priest. The same thing happened when she predicted her uncle's death.

Helena didn't know how to relate to people. Only Phillipe supported her; he was her only company, the lone person with whom she could share her thoughts and worries. This acceptance had helped Phillipe open his own mind and sometimes he was able to intuit things like her.

After the experience at the beach in Spain, Helena had been struck by the idea that we live and die many times. She wondered if it was possible to remember these lifetimes and what would be the reason for living so many times. She began to distance herself from the religious ideas that her parents had imposed upon her to delve into her inner world; she wanted to find answers. Little by little, time passed and her small world changed. With the help of Phillipe, she spent time reading about other cultures and ideologies that saw the world in a different way than she was accustomed to.

"Maybe, if we can remember our past lives, we'll understand why we exist," she said to Phillipe on one of those afternoons when they hid from their parents and shut themselves in her room to read. The things that she had been taught at home and in school, and which she had previously accepted without questioning, became things to be wary of.

Helena observed the photograph hanging on the wall in her kitchen in Remoulins. It was one of the pictures that Phillipe had given her and the only one she had kept. She remembered her brother's excitement when he had bought his camera, a Nikon 35 mm, with the money he had saved from his summer jobs. He took it everywhere, disappearing for hours while he explored the city, searching for unusual details in both the people and the interesting buildings he encountered. It was he who had taught her photography. Whenever he could, he would take Helena with him, explaining to her about lighting, angles, composition. Helena had taken pleasure in these outings with him, enjoying her time out of the house, and had taken up photography because of her brother's enthusiasm.

Now, seated at her grandfather's table, Helena smiled at this memory of her brother, always happy and ready for any adventure. How she had missed him when he had died. She was barely fourteen and was left on her own. Without Phillipe, Helena had felt adrift. To protect herself she had learned to hide who she was, adapting to the world, trying to find a way to fit in so that people would accept her. She hid her abilities and never talked about them. She would shut herself in for hours to read and draw; at least that way she was able to express a little of what she was feeling. She had settled in at school and focused on her studies, finding enjoyment in learning about science and history. However, she knew deep down this wasn't what she was searching for. She still hoped to find someone who could help her understand her restlessness and find the answers to her questions. She had thought, 'there must be something else, something that gives meaning to our existence. I want to know where we're going. I suspect that everything is made with a purpose. Who can tell me the truth?'

After her brother's death, Helena, anxious to escape the oppressiveness that she felt, had found refuge at the farm of her grandfather Adolfus. He lived alone in a small stone house outside of the town of Remoulins, where he dedicated himself to caring for the olive trees and vegetables that he had sown. There, during her frequent visits, Helena rambled for hours in the hills, walking to the river and observing the animals that lived in the area. While she meandered she recalled Adolfus' chats about nature and the animals and how he was teaching her to care for the garden in his house. She always took Phillipe's camera on these hikes, and so it was that she began to photograph the landscape and the olive grove, the clouds, the neighbors, anything that she found interesting.

The introspection that she lived in those years had helped her to hear what in that moment she had called her intuition. It had presented itself in various ways: dreams, premonitions, sensations, thoughts, or as an inner voice. When she had attuned herself to her inner "guide," her perception was more defined, more precise.

And so the years had passed. She had felt as though she lived behind a mask, never showing what she really had inside. She continued to take photographs and it provided her with the means to get out of the house. Her parents, seeing her more adjusted, had given her a bit more freedom and let her wander within the old city walls of Avignon. She took advantage of this free time and made the rounds of the city on her bicycle. She lived all of this feeling as though she were on hold, waiting for something to happen. When she was 17, she decided to study medicine. She was very sensitive to the pain of others and thought that by helping to cure people she would find a purpose to her life. She was resolute; she would go to Paris to study. The city attracted her greatly and she knew in her bones that it was important that she go there to live alone. She worked diligently at her studies and took the admissions test for Pierre and Marie Curie University.

She had done all of the paperwork clandestinely so her parents wouldn't find out. They maintained the position that they didn't want Helena to study; they saw no purpose in it, and less still that she would be so far from home. They still had hope that one day she would choose a more traditional role and get married. Why bother putting so much effort into a career if one day she was going to leave it to have a family? And above all, one that was so difficult to achieve.

Her father always pointed out to her, "Only fifteen percent of first-year students make it to the second year. What makes you think you can succeed?"

Once in a while her mother tried to get closer to her and made an attempt to listen, but Helena broke with all of her notions of how things should be; she frightened her mother and only earned her mistrust.

'Why do we always react with mistrust or aggression in the face of something different?' she had asked herself. She had realized that ideas that fall outside of the norm are rejected because of the fear that change produces. She had said to herself, 'we know that we live in an illusion; deep inside of us there's a little voice that tells us so. It warns us that the foundation of the building that we have constructed painstakingly for so many centuries is not based on anything real. Human beings are predisposed to react with aggression and negation in order to defend the illusion because we know that at the slightest touch the false idols of homeland, family, and religion will come crashing down.'

One rainy afternoon, after weeks of waiting, she received the longed-for news: she had been accepted at the medical school in Paris. She was thrilled; at last a door had opened before her. But now she saw various obstacles. 'How am I going to be able to live alone in Paris? How will I be able to pay my expenses?' She had spent the afternoon in front of the bedroom window watching the rain fall. She felt trapped; she yearned to go but at the same time felt fearful of facing these new challenges. And on top of it all, she'd have to confront her parents. That afternoon had seemed eternal. Standing in front of the window, she watched as one raindrop fell on the glass, observing how it followed its path until it found another and joined with it, growing larger and then uniting with others, until it achieved a force that made it autonomous and was able to slide off, whole and complete. That was how she imagined herself; she would have to gather the strength of all of the experiences in her lifetime that had impelled her to move forward.

She had said to herself, 'I know that I need to go to Paris and be alone there, and not just because of school. I know that I'm going to find something there.' She noted a great force inside of her that compelled her to continue, to not stop herself. Her inner voice told her that she needn't worry; she just had to take the first step.

The next morning she had taken the bus to go visit her grandfather Adolfus. She didn't know if she could expect anything from him, but she needed to talk to someone. Even though he was old, he had a young spirit, younger than that of her parents. He had adapted better than they to the changes of the 60's and was pleased by Helena's strength and her efforts to live in a different way.

Helena's parents, even though young when these changes had precipitated all over the world, belonged to a group of people who fear new things and try to hold onto what they have, not wanting their world to alter. But when Helena explained her situation to Adolfus, he offered to help her with part of his pension.

"I can't give you much; you know that I spend more than I make here at the farm, but I'll be happy to help you for a while until you get installed in Paris. After that, you'll have to see what you can do on your own." He watched her for a bit and then continued. "Maybe you can get some good use out of that camera that you take everywhere." And so began the second stage of her life.

In the autumn of 1986 Helena enrolled in the University in Paris. She hadn't asked her parents' permission, nor allowed them to say anything to her about it; the matter was more than resolved. Neither did she reproach them anything; everyone lived what they had chosen for themselves and each should be responsible for their actions.

From the start Helena had savored the independence that she found at the University. She relished going out to explore the city in her scarce free time and she amused herself taking pictures. She found a satisfaction in the first three years of her studies and then in the years of internship, where she had dedicated herself completely to her training, utilizing her intuition to diagnose patients in a very accurate way. The other students were sometimes amazed at the precision with which she identified the cause of their illnesses.

Helena lived together with her classmates and they got along well. They enjoyed her conversation, which in addition to being different was peppered with humor and irony. Nevertheless, she felt as though she didn't share their interests. They were looking to make great discoveries or leave their mark in the world. Others were merely fulfilling the expectation of their families and still others just wanted a comfortable life and a private practice. She longed for something else. She often roamed the streets alone, searching without knowing what she was looking for.

When she was 23, Helena was weeks away from finishing her internship at the University. A few months earlier her old restiveness had returned. She had had various dreams, filled with images and symbols which made her once again feel the emptiness which she had buried inside her while she was so occupied with her courses. In one of them she saw herself in a city. As was her custom, she wrote it down in the notebook that she had by her bedside:

> I see myself walking through the streets of a city; I don't think it's Paris because there are many buildings which I've never seen here. I'm accompanied by several people, some of whom I know. But little by little I leave them behind,

as they become distracted in the shops or doing other things. I stop in front of a building that I don't know. I enter through the front door and inside there are lots of people and children, I hear the voices of various persons shouting and there's music thundering in the background. I see a large spiral staircase and walk up it, using my foot to push aside the things that are blocking my way. I arrive at a large door, open it and enter. I'm amazed to see the contrast between the filth outside and the cleanliness inside. I enter a room that is bathed in a golden light that impregnates everything. At first there's no one there, but in the background I see a woman who approaches me. She's tall and strong and she invites me to come closer with her gaze. Without speaking she asks me to give her the ring I'm wearing. I extend my hand and discover that I'm wearing the same ring that I saw in my childhood vision, but now it's dirty and battered, and the stone doesn't shine. I take it off and hand it to her. The woman places it in a flask with clear water. After a moment she takes it out and places it in my hand. The ring is clean again and shines with a golden radiance. I can't believe that I've seen that ring again after all these years. What can it mean?

That afternoon she had gone out to walk in the Latin Quarter. Passing in front of one of the cafes that abound in the area, she heard someone call her name. When she turned, she saw one of her classmates from the University, who invited her to come over. Helena walked over to the little table where her friend was chatting with a woman who immediately captured Helena's attention. When she saw her eyes, she recognized the woman from her dream. Her friend introduced them.

"Well, you finally arrived," were the words that Marion said to Helena. "Have a seat and join us for coffee."

Intrigued by the manner in which she had been greeted, and wanting to know why she had dreamed of her, Helena accepted the invitation. Marion was approximately 55 years old. She exuded a great vitality that seemed to fill the room. She possessed a gaze that seemed to read the thoughts and feelings that Helena had in that moment. Her gestures and movements were graceful and exact. Helena noted that she emitted a faint fragrance of roses. She felt wrapped up in the energy that Marion radiated. She made her feel small and exposed and she felt that there was nothing she could hide from her. At the same time that she felt a comforting closeness, she perceived an enormous distance that separated them, as if they lived in two distinct worlds.

In that instant, Helena's friend excused herself because her boyfriend had just arrived to pick her up. Marion and Helena ordered espresso and cake. Seated alone at the table, they began to talk. The hours passed and Helena was shocked to see that it was ten in the evening when they parted. She walked for a couple of hours after the encounter trying to assimilate what she had just experienced.

Upon arriving at her apartment, she opened her notebook and jotted down the details of the meeting.

Sitting alone with Marion at the table I felt exposed, as though she could read my thoughts and emotions. She treated me with familiarity, as if she knew me. For my part, I wanted to know why I had dreamed of her. I was intrigued to find out who she was. She asked me with great gentleness why I had such a melancholy gaze and what it was that I missed. This comment awakened many memories and sensations in me that I hadn't felt in a long time; it was as if something had broken inside me. I could only answer that I had always felt a great emptiness, but I've never known where it comes from and it's accompanied me since childhood. She explained *to me that when a human being has had a spiritual development, they can no longer find contentment within human reasons, because they have glimpsed the true reasons that are within them. She told me that humans suffer from a great imbalance; we have fear that we project in myriad ways, such as anger and sadness, among others, because we aren't living according to the reasons that correspond to us. She told me that we have lost the consciousness of who we are and what we need to do, but all of this can be overcome if one studies.*

She told me the story of the beginnings of man, of why we live in an illusion and not in a reality. She spoke to me of great reasons of the Universe and of Life. Hearing her speak of all these things, I felt as if something inside of me aligned itself.

Time moved so quickly, I didn't realize it in the moment, but we were there for several hours seated in the cafe. She told me that she had to go and implied that she would be in the same spot in a couple of days. When she left, I felt like a cloud floating in the immensity of space. I'm absolutely certain that I've found the person who will teach me.

In the two days that followed, Helena had tried to finish the application for her medical residency. She had to decide what area to specialize in and the hospital she would apply to, but she couldn't concentrate. Questions that she wanted to ask Marion kept filling her mind.

The third day arrived and Helena showed up early at the cafe, thinking that she would get there before Marion, but to her surprise she found her already seated drinking coffee and smoking a cigarette. They greeted each other and Helena had barely settled in when Marion began to speak.

"What do you think of Paris?" she asked. "What have you perceived of the city?" Without awaiting a reply, Marion continued. "Don't you think there's a reason for which ground-breaking artists, writers and musicians have come here throughout the centuries? In the Earth there are different locations which receive a greater charge of energy than others, which nourish man's thoughts. We might say that the Earth has 'veins' of energy that run through it and there are places in which this force is more perceptible. Energy presents itself in these spots for a certain amount of time and then it moves to another in concert with the reasons of the Earth. We look for the places that have a compatible energy when we have a deficiency. *The Earth has, just like the human body, specific sites of power. [...]As the human body has channels through which clean blood courses, and others for unclean blood, thusly in the Planicio Terráqueo (Earth) there are sites which thou mayest call adequate and inadequate. The inadequate places, for the pequeños (little ones) [...], are highly negative as they absorb their energy and vitality. Conversely, the adequate places nourish their presence (29 Sep. 00).* These places that nurture will be healing centers in the future." Marion paused, noticing the question forming itself in Helena's mind and continued. "Why is it that artists, for example, are more receptive to this energy? Some of them have gained a certain elevation outside of the continuum of human thought. Elevation means to distance oneself from human reasons to approach divine reasons. Now, the new word we should use instead of divine is 'Kosmic.' For the moment, I'll just say that this is what lies beyond our human understanding."

Marion spoke of the painters that she most enjoyed, of their techniques, of how art had transformed the ideals of human beings. Helena was fascinated listening to the wealth of stories that she knew about the painters, the range of colors, and methods that they had used. It seemed that Marion knew all about it. She spoke of how art reflected the search that had existed in each era. She explained how the great artists had suffered the rejection of their contemporaries because in order to express something novel, they had had to open themselves to be able to glimpse something beyond what was already known. In this way they had been able to break with a fixed structure in man's mind, and thus, in themselves. She said that artists didn't create or invent anything new; they simply took up something that was already manifested by the Supreme Entities that propel and foster everything. As an example of this, Marion explained to her that the Universe is like a great library full of books. Sometimes, in specific moments, our konsciousness opens and we are able to see a book that we hadn't perceived before because it was beyond our capacity.

Helena was intrigued when Marion recounted that it was necessary to search, to struggle, and many times suffer in order to be able to step out of the continuum of human thought and be able to reach the wealth that is beyond the mind of man. She was amazed to hear that humans don't perceive form or color in its totality because they are foreign to their basic structure and **only painters and sculptors preserve this relationship of perception of form and color (Carpeta Dorada).**

Even though all that Marion told her captured Helena's attention, she was waiting for the moment in which she would be able to ask her about the things that she had recounted to her in their first meeting.

However, she intuited that for some reason she should wait until invited to speak.

When Helena was totally absorbed in the conversation, Marion looked in her eyes and said, "When a pequeño, or little one, as the Konceptos (Koncepts) call us, searches for the Konocimiento (Knowledge), he is taken into account by the Supreme Entities that manifested him and he is granted an opportunity. The pequeño must decide in that moment if he fully accepts or not. If he accepts, he must fulfill each of the tests that are asked of him. What are these tests? To put aside that which disturbs his path. ***In order for understanding to rule within us we must obey, but analyze the cause, so to avoid falling into the submission of a robot that doesn't arrive at the important objective, which is to comprehend" (21 May 88).***

"And what is the Konocimiento?" Helena asked her.

"The Konocimiento (Knowledge) is for you, what you understand as the beginning of everything (20 May 95).

The Konocimiento is a primogenea (primary) reason, a multiple of reasons and causes that are motivated within themselves, and that remain in great changes, as a never-ending process (5 Oct. 93).

The Konocimiento projects itself in the creative fields and it is not till now that we have understood that what we see, what we feel, are nothing more than projections of the Konocimiento (27 Nov. 93).

The Konocimiento is not an action or a fact [...] but it manifests itself within actions and facts, within Life (5 Oct. 93).

In this HTime, the Konocimiento Kósmico (Kosmic Knowledge) has been delivered to us. In the past, humans received the teachings that we know today as Sabiduría (Wisdom). The Konocimiento Kósmico is not within the mind of man, it's something new to him. Wisdom is an appendix of the Konocimiento (Knowledge). What I'm going to talk to you about are new teachings that surpass all that mankind has. In order to receive it, you have to deserve it, you have to earn it." Marion looked at her fixedly and continued. "You need to know that the restlessness and uneasiness that have accompanied you in this life are because you've had several lifetimes of preparation. This disquiet will end only when you once again integrate yourself into your path. There exists a continuity within us that registers all that we have learned throughout our incarnations. A person who had a progress in the past carries this advance inside of them, even if they are not conscious of it. Due to this advance they no longer find satisfaction in the reasons of the world. It's because of this that since childhood you felt alienated from the world you live in. To aid you, you were presented with visions and telepathy, so that you could glimpse something beyond the ordinary. This is something that you developed in the past and you can take advantage of it now to help you in understanding your past. You should put your efforts into reconciling the reasons of times past with those of the present so that you don't lose the continuity once more."

33

Helena felt relief at being able to understand something about herself, but at the same time she felt overwhelmed to hear that she needed to remember her past.

Marion had continued, "Mankind, and above all young people, feel a great discontent with life. Deep down they know that they've lived the same things many times; we can say that within the kaleidoscope of life, they have been through all of the possible experiences and emotions, and don't find satisfaction in anything. They don't know where to go. And the force of the continuum of human thought is so great that it envelops them and doesn't allow them to glimpse other reasons, either due to their own fears or apathy. Sometimes they end up mixed up in drugs, or in sexual depravity, looking for strong emotions that allow them to break out of their stupor, or they surrender their energy to the self-same continuum."

Helena continued thinking for a while about the few memories of other incarnations that she had glimpsed in her life. Since the vision in Spain, she had dreamed of other moments from her past, but they were loose pieces that she hadn't been able to connect. She asked, confused, "But how can I remember the past?"

"This is the process of knowing yourself, of understanding who you truly are. I'm not talking about your habits or tastes. You're not what you think you are; this is only a temporary appearance linked to your current body. *The Ego is semi-induced in an artifact or vehicle with which it roves about and in which it lodges its particles in the brain, and thus, operates it. It is its own material body (5 Sep. 87).* We have to think for ourselves, be individuals, and thus we begin to know ourselves. Look, *man does not learn to think because everything is handed to him already thought out, not leaving him the chance to think for himself, of what he wants, and understand. He is in school for twelve years, saturated with useless ideas (at 12 years old the konsciousness of man awakens and at 21 he already has a libre albedrio [free will]) and he is not taught to know how to think about what he wants. Man must know himself first and foremost. This will lead him to know what it is that he really wants in life (4 Nov. 87).* But look what time it is. I should go," said Marion, rising from the table. She picked up her purse and her shawl and bid her goodbye, saying that she would see her in the same spot in two days. Helena left the cafe overwhelmed. She was curious to know more, but also recognized the weight of each one of Marion's words and the responsibility that she had expressed.

The following day, Helena awakened in a quandary. It turned out that the day she was to see Marion again was the same day that she needed to take the exam for her medical residency. If she didn't take it, she would lose the chance to continue in her studies. After so many years of hard work, she didn't want to leave them. But she also knew that if she didn't go to meet with Marion that she might never have the opportunity to see her again. She had so many questions in her mind; it was as though all the concerns that she had guarded inside for so long were reappearing at the same time.

The two days were tumultuous; Helena spent them trying to decide what to do. She had a dream in which she saw herself in the middle of an immense sea, alone and adrift; she was drowning. When she surrendered, she saw a boat approach her and strong arms pulled her out of the water. She awoke with a suffocating sensation. She remembered what Marion had told her, "The opportunity only presents itself once."

Helena hadn't understood it in that moment, but some months later Marion had explained to her that in moments of great transcendence there is a continuous struggle between the two forces within us. "Within us are different levels of konsciousness. *To find the deepest, we must enter into the reasons of the Lux (Light) that are in us. [...] There is a konscience that rules us, vital, that makes us comprehend: I want to go beyond, I want to evolve, but in our inner konscience there is a programming against this. [...]The internal konsciousness still impels us to rebellion*" (21 May 88).

On the afternoon of the third day, Helena had left for the encounter with her destiny: she let herself be guided once more by her inner voice. She arrived at the cafe at the same time as Marion, who, upon seeing her, had taken her gently by the arm, saying, "It's a lovely day for a walk, don't you think?"

That day, Marion was dressed in a very elegant manner, with her hair arranged stylishly. She was in a lighthearted mood and gave off an air of such vitality and energy that in that moment Helena had forgotten her worries and ruminations. It seemed to her that she entered into another dimension when she was with Marion. They had walked for a while when Marion said to her, "Look, we're very close to my studio. Why don't we go up so you can see it?"

She lived in a three-story building that had been constructed at the end of the nineteenth century. The owners had converted it into lofts for artists a few years earlier. Marion had chosen the place because the apartments, although old, had enormous windows through which entered a wonderful light. They walked up the stairs from the street and passed into the simple foyer. They got into the noisy antique elevator and when they arrived at the third floor, they entered Marion's small apartment. Although the space was simple, it was nicely arranged and very welcoming. There was little furniture, but the plants were placed in pots of fine Chinese porcelain. A large table and a wooden chair in the center of the room dominated the space. The table was covered with an abundance of papers and notebooks. In front of the window that faced the street there were two armchairs upholstered in red leather and a glass table between them. A Moroccan-style rug in earth tones covered the oak floor beneath the chairs. The walls were painted white, with the exception of the one in the small kitchen which was painted pomegranate red. Propped against all of the walls were many canvases; none of them was signed. Now Helena understood why Marion had talked in such detail about painting. Even though she had been invited into the studio, Helena didn't feel comfortable looking at the artworks as much as she would have liked. She felt that they were something very personal and intimate. Marion invited her to sit while she went into the kitchen to prepare coffee, but Helena followed her, offering to help. They prepared two Turkish coffees in an old bronze coffee urn. They sat in the living room and Marion lit a cigarette, saying,

"Now, tell me Helena, what are those questions that you've been so anxious to ask?

"What is the purpose of life? I'd like to know why we live many times and where are we going?" Helena reeled off a list of questions.

Marion smiled at seeing Helena's mettle. She liked hearing questions that were different from those that she was accustomed to hearing, like, 'will I be happy in my life?' or 'will my business do well?'

She responded, "Life is a module of learning. Each lifetime is a lesson that forms part of a continual transformation that should lead us to an understanding and an advancement. The material body that we have is simply a disguise that we change in each new lifetime. *Man is ephemeral, but the Ego that animates him is forever" (Nov. 10).* She saw the confusion on Helena's face and continued to explain, "The Ego, or more accurately, the Yo Ego Deidad, is the Real part of us. It's not the same ego that psychologists talk about. *The Ego is a Kosmic Atom with a distinction, we have a part that is Real which comprehends and says 'I exist', 'I am' (5 Aug. 06).* This Kosmic Atom forms part of the Universe or Fusion in which we live. The Ego doesn't die; it's the part of us that is in a continual learning process. The matter that we have on the Earth is a temporary appearance. We Egos pass through a series of transformations so that we learn, comprehend, and evolve. Evolution is understanding and advancing in order to then move on to another reason. For the time being it's not easy to understand the Ego, since we're so involved with the material body and all of its demands and needs; it's our matter which usually governs our actions. In order to embark upon these studies, it is vital that we differentiate between the needs of the Ego and those of our matter. *We should begin by understanding that what we appreciate of man, that is, the material body, is nothing more than a disguise imposed upon the Ego, and that this disguise changes in each incarnation in order to give the Ego a motive for understanding and analysis, which is part of the lessons that the Ego must have on the Earth" (27 Jan. 05).*

"And what is the meaning of life? Where are we going?" Helena asked her.

"For now, the first step is to understand why we are on the Earth and why we were placed in a body that we didn't have before. We are here to learn. Learn what? Our lack of humility and submission has distanced us from our path and our place in the Universe. The time on Earth is to overcome this shortcoming. We have to pass through Seven HTimes here in the Planicio Terráqueo or Earth, each one with specific teachings and conditions in order to be able to reach wholeness again. For that reason it's not permitted to leave the Planicio Terráqueo before completing our lessons; nevertheless, mankind is currently trying to do it with spaceships. When we have understood and completed the requirements of all of the HTimes that are programmed, we will once again have a more subtle matter that will have the form of a flame. We'll have the graces that we had before and we'll return to complete what was left pending in our first home, the Kosmo of the Trinak*." (*This is a place outside of the Planicio Terráqueo or Earth where we lived before, and there we didn't have a material body.)

The questions collided with each other in her mind. Everything that Marion was telling her was new and she struggled to assimilate and comprehend it.

"Why do you refer to humility and submission?" asked Helena.

Marion responded, "They're vital if we're to accept that we were made with a purpose that we have to fulfill. Our behavior must be consistent with this purpose or reason and we should only take what is permitted to us. If we were prepared to be cooks, why do we want to be bricklayers?" Marion chuckled and continued, "The teachings of which I'll speak to you are novel and have never before been given to us; it is the Konocimiento Kósmico (Kosmic Knowledge) of this Third HTime. Up to this point mankind has received Wisdom, which is a constituent of the Konocimiento, and with it he has modified and mutated his world. The Konocimiento Kósmico will allow us to find us our place within the Kosmo that we belong so that we can understand ourselves as a part of it. Our development and evolution will only be achieved through our own efforts. The Ego that is initiated and committed to its evolution is granted the Grace and the permission to be able to grasp the Supreme Entities that exist in the Fusion in a direct way. You'll see, Helena, life in this sphere has as its purpose an apprenticeship, and it is by means of this that all Egos will be able to remember who they are, where they come from and where they need to go to. It is a labor that corresponds to all, according to the level of konsciousness and development of each. Every one of the

Egos has a reason to fulfill in each lifetime; if they achieve it, they will move closer and closer to their origin."

Marion had remained silent while she lit a cigarette. She inhaled deeply and looking at Helena, she asked, "What will you do? Are you going to accept your opportunity?"

Helena had been waiting for this moment for centuries, although she still didn't know it. "Yes, I accept. I want to learn," she answered, longing to be able to understand all that Marion had told her.

Marion marked this step saying to Helena, "At this moment you are recognized before the Konceptos (Koncepts) so that they may guide you and teach you; let there be respect and humility in you to be able to accept them. He who is initiated in this path of evolution receives the help of the Konceptos."

"What are the Konceptos?" Helena asked her.

"They are excellences of the Konocimiento itself that deliver and direct the motives of development to all of the particles that are in evolution in the Universe, or as we say, the Fusion," she responded. "They allow us to be aware of and relate to the Konocimiento. There are Konceptos that humans can perceive, since they are 'closer' to our level of konsciousness. When we elevate ourselves by means of study, we may, with their permission, be able to be aware of them. There are others that, due to their high investiture, are beyond our meager understanding.

The Konceptos (Koncepts) are aware of all that is, and all that we perceive*, and thus, they perceive what and how is the development of the forms of life in the multi-dimension. For the Konceptos the created, the projected, is like a 'language,' an effective 'medium' that serves to ascertain the development of what there resides, and which are no more than projections of something we are incapable of comprehending, but rather, only through them" (27 Nov. 93). (*That which "is" is All that exists, it is Real, even though we are not able to perceive it. What we perceive is Fantasy; it is what we are aware of within the limitations of our current konsciousness in the third dimension.)

For Helena, this had been her second birth in this lifetime. She had found what she had been looking for since childhood and long before that, but as her Maestra Marion had said, "Your road is just beginning." Her Maestra had bestowed upon her the possibility of kommunicating with the Supreme Entities that constitute us. Although, in order to realize this enormously transcendent step, she would need to dislodge herself from her personal interests and find the submission to allow herself to be guided.

The following months were filled with huge changes for Helena. She had finished her internship, living a little while longer amongst her classmates, until the moment had arrived when she needed to begin her residency. She decided not to go. She said nothing to her friends, nor to her parents, because she knew that they would never understand. This had been the most difficult decision for her; the University had been her joy and her anchor in the world. She was letting go of what she most cherished. But in order to understand the new reasons that she was going to study, she had to begin to see the world in another way. Because of this, she could no longer read about the science that she had so loved. She had to distance herself completely from that environment.

In order to be able to stay in Paris, she had gotten a job as a waitress and kitchen assistant in a cafe. Later on she had also begun to take photos for an advertising agency, and in this way she was able to meet her expenses.

To help Helena understand the necessity of these changes, her Maestra had given her an example.

"Our mind is like a glass that's filled with the thoughts and habits which we have repeated for centuries and which form the continuum of human thought. It's necessary to cleanse ourselves of them, to empty the glass completely so that it can be filled with new reasons and thus these won't be muddled with the old ones."

So it was that Helena had taken the first steps in a road of learning that would lead her to the encounter with her true self and to know the reasons that form the Ego, finding in this for the first time the contentment of feeling herself part of All.

Now, seated in her cottage in the middle of the old olive grove, Helena beheld herself with the distance that comes with the passing of the years. She looked within for her connection with the Supreme Entities and gave thanks for the opportunity that she had been given, wishing that this same possibility could arrive for all. She knew in this moment that she needed to tell the story of humankind since before its arrival on Earth, that tale that had been forgotten in the course of time and which humanity needed to be aware of. Recognizing these reasons, mankind would be able to accept the supreme gift of the Konocimiento.

Chapter 2
From Helena's Book:
Who are we?

Eager to speak of the truths of mankind, Helena began to pen her book, recounting the history that had been relegated to the far reaches of the human mind. Fulfilling what the Konceptos had signaled to her as her responsibility and esteeming what the Konocimiento explains about the reasons of our existence, she wrote:

Throughout the ages, the great thinkers and scientists have wondered, "Who are we?" and even though they have searched for a meaning to the reason of our existence, they have not arrived at the truths that we can find in the Konocimiento (Knowledge) that has been given to us in this Third HTime. In order to comprehend who we are, we need to understand that the material casing is transitory and ephemeral. *The physical appearance is like a little peel, like that of a fruit. This is temporary and projected (05).* We only know ourselves as human beings, and although we believe that we understand the matter that envelops us, we've forgotten what our true nature is. This truth has been buried in the long centuries that we have been here on the Earth.

What are we? *[...]We are Egos. This is our Vital Principle which will endure in us forever (14 Mar. 11). Ego signifies that which develops in principle and forms part of the Voluntad Suprema (Supreme Will) (Carpeta Dorada). The Ego is the integral part that constitutes us absolutely (10 Aug. 10).*

The Ego is the Kosmic particle that in each incarnation or Ansibir enters into a new body. It is separated from our matter when we Trebolar (when we enter into the state of Trebolo, or what we know as death) in order to undergo a preparation and then enters again into a new matter in which it will have different experiences and lessons, with their joys and their difficulties. Therefore, we don't die; on the contrary, we are in a constant transformation.

When we speak of our beginning, we should say that the Ego was not born, but rather it was manifested by a Great Excellence that we are unable to comprehend with our present konsciousness. *The manifestation is in itself, existence (28 Aug. 10).*

If we ask, "Why do we exist?" the Konocimiento explains to us that *existence is an invitation to a grand adventure, which is that of being accepted in a great creation (28 Aug. 10).* We form a part of all that is found in the Universe, as if it were an immense family.

The Ego, after being manifested (when we can say that it exists) continues in a "preparation" *[...] in which for a given period it is granted education, comprehension, and rights and obligations in order to act within the Great Creative Work; after these teachings, which are multiple and vitalize the Ego explaining to it step by step and fundamentally suffusing it with what it will be, it is conceded a decision of "I exist or I do not exist." This implies a commitment in which it is granted its wish. If the Ego accepts the obligatory reasons, it passes to another reason called* PROJECTED, *in which it will fulfill itself within a Real and an Aparente (Apparent) and will have a specific function according to the realizations of the Creative Fusion (28 Aug. 10).*

Into each Ego was placed a particle of will, of amazement, of cleanliness, of candor, and of instruction that registers, simplifies and activates the environmental teachings (Carpeta Dorada). Each one was prepared to be able to fulfill a purpose or reason. Within the process of our conformation we acquired the specific characteristics for that purpose. *The Egos were made through the coexistence of the Protentos or Supreme Entities, even though in the same way that a cake is primarily flour, we belong more to the reasons of one of them, the Koncepto Lux (Light), than to those of the others (19 Sep. 87). [...]Our formation was a structure of light (acoustic or sonorous reasons that are part and equilibrium of the Lux) (21 May 88). Each Ego has a specific sound, (in the present, this is related to the personality) and no two Egos are alike (Carpeta Dorada).* Being formed principally of Sonorous Lux (Light), everything related to the Ego has to do with sound: our perception, projection, konsciousness and thought.

The Egos were formed in sonorous groups or Huestes (Hosts) of different levels according to the sounds and purpose of realization that we have. Even now, this way of grouping ourselves is still valid; it presents itself in the affinities or rejections that we feel for other Egos or groups of Egos, in the reasons we share, and in the lessons that we need to understand in order to grow.

But we return to the reasons of our beginning. When the Huestes of Egos had been formed and prepared to fulfill their missions, the great moment arrived when they would be projected into the Universe, or as it is called in the studies: Fusion, Campos Experimentales (Experimental Fields), or Expansión Creativa (Creative Expansion), and thus began our development and experience within the principle of Life.

And what is Life? *Life, as thou sayest, is an action of the Konocimiento. [...]The Ego is a realization that belongs to*

Life. When it enters into Life it belongs to creation. Within creation, to the actions projected within the Projection (1 Mar. 94).

We understand the process of life on the Earth as the cycle of birth, growth, reproduction, and death. However, the Ego doesn't die; it transforms itself constantly. Life in the Fusion is a process of continual learning and transformation that leads a particle, in this case the Ego, to evolve, that is, to pass through a development that must result in an improvement and an elevation. Upon elevating itself, the particle comprehends more and more and approaches, by means of this comprehension, the reasons of its origin and purpose. An Ego must seek its evolution in order to perfect itself. How? By following the Voluntad (Will) of the Supreme Entities that formed it, because this Voluntad leads it to fulfill the reasons of its existence. In this way the Ego will Realize itself and will pass through multiple experiences and transformations that are not possible to understand at this time. If it doesn't follow this Voluntad its progress will deviate from the planned destination and from the reasons of its existence.

The Fusion in which the Egos were projected is like an experimental laboratory where all of the reasons projected within it are in a continuous apprenticeship. The planets, the galaxies, the atoms, the Kosmic workers, the sun, and all of the other reasons that are found in it, direct themselves towards their own evolution.

We see that within the Fusion there are many Kosmos; each is a host of reasons that follow their appropriate laws of development. The Egos, formed and prepared to enter into development, were projected to the Kosmo del Trinak (Kosmo of the Trinak), our first home, which is a Kosmo of sound. And there *we should have belonged in wholeness and grace to fulfill a mission in preparation (Carpeta Dorada).*

In the Kosmo of the Trinak, or "Trino" (Sound) we Egos began our development. We carried out our duties in "constructing" sonorous forms. *A sonorous Ego belongs to the reasons of Grace. It is utilized in the affirmation of a Creation (23 Aug. 90).* In that Kosmo, the Ego did not have matter like it does now. *It was thinking sound that understood, moved, and communicated (29 Mar. 89).*

For a time we were in harmony with our reasons, with the Supreme Voluntad (Will), and with the Konceptos; so what happened to change that? Why were we projected to the Earth, or Planicio Terráqueo, that we presently inhabit?

This is the history that we must understand in order to begin upon the road that leads to our destination.

Helena closed her notebook, thinking of the efforts that her Maestra had made throughout so many years in order to help mankind advance and fulfill the reasons of its being. Grateful for all that Marion had taught her, and for the opportunity that the Konceptos had given her, she rose, took her jacket and headed out to hike in the countryside, enjoying the last mild days of fall.

46

Chapter 3
The First Steps

The old refrain says:
"Take care of your home and not your neighbor's."
Mankind doesn't take care of its home,
but wants to take care of the lands of the Universe.

AUDACITY

Strolling through the streets of Remoulins, the autumn light awakened in Helena the memory of long walks along the avenues of Paris in the company of her Maestra Marion. Although eighteen years had passed, she remembered it vividly: the small details of the places they had visited and the first talks they had had. In those initial meetings, Helena had been astonished in the face of many of the things her Maestra had told her. At one point she had commented to Helena, ***"Man is a re-creator of forms, thoughts and reasons; everything is already made and he absorbs it all in the recesses of his brain and his heart. The act of creation does not belong to him, but is an undefined loan" (Nov. 10).***

For Helena, this simple phrase had overturned her acceptance of what she had learned over the course of her lifetime: that human beings had discovered, invented, or "created" their environment. She couldn't but feel small and humble in the face of this reason, understanding that we only gather in a part of what is already formed and projected by the Konocimiento (Knowledge). She had thought, 'it's really as if our konsciousness hadn't changed since we thought that the planets and the stars revolved around the Earth. We still see ourselves as the center of the Universe and as such, with the right to do what we want with it.' She wondered, 'would we feel humility if we knew that we are the only ones in the Fusion that are limited in an elemental matter?'

Helena had grasped the truth in what her Maestra told her, even though at first it was hard for her to accept certain things without trying to reason. Like the time when she had said that man had been living for millions of centuries on the Earth; Helena hadn't understood how it could be. But Marion had explained to her, "Time isn't like man perceives it now. Sometimes millions of years are spoken of, but the Ego didn't feel it as such. In the studies we see *[...]that the presence of the Ego in the Planicio Terráqueo (Earth) was and is of thirty million centuries, but we have to declare that the times per se were not equal amongst themselves, as they suffer modifications in regards to the Kosmic events that the Fusion represents, and experience an adjustment in relation to the self-same Planicio Terráqueo.*

Ye know that HTiempo (HTime) has very diverse variants, being itself only one, and that it projects itself in a multiple incomprehensible for ye. Because of this, ye will understand that the Ego, throughout the ages, took vastly variable forms, but in each there was always a Konceptual point, that which now ye call and know as konsciousness. The forms that have governed the Ego are very variable and have been modified since the beginning, like the cell, until your days, in which ye have taken a form which ye call human.

We will continue in a constant transformation. *Your material presence of this moment will be very different in the following centuries. The previous civilizations, as ye now say, were not always equal. In some, the pequeños (little ones) lived 30 or 40 years, and in others, 300 or 400 years. This, it must be made clear, according to the measure of time in which ye understand it"* (23 May 01).

Marion had continued, "The unfolding of the Planicio Terráqueo (Earth) and all of its reasons are already formed within the Seven HTimes and will go on displaying themselves in the course of the centuries to come. It's like a book already written, and when a page is turned, the new reasons become 'visible,' manifesting themselves in the changes in the Earth, in the plant kingdom, in the animals, in our matter and principally in our konsciousness. Remember that we are sonorous; *we know that for the Ego that inhabits the Planicio*

Terráqueo, whether in the stone, in the animal or as a human [...] life is a sound, since the Planicio Terráqueo is a site especially designed so that sound reproduces itself in it (10 Aug. 92).

Our progress is integrated and coordinated with the progress of the Planicio Terráqueo. Between them exists a connection that until now we haven't understood. *Furthermore we should understand that it is the different energies that are in the Planicio Terráqueo which vitalize or nullify part of this process of development of the Ego, since there should exist a harmony between the Ego and the Planicio Terráqueo itself, as this also has its own development (27 Jan. 05).*

What thou seest then as the complexity of thy body has been instead the development of thy own konsciousness. Thy development has been then part of a plan but thou hast participated in the same.

In this thou shouldst understand that thy development and that of the Planicio Terráqueo itself, is like a great orchestration in which the vital principles, as it were, of life itself, favor each other until they reach plenitude (18 Feb. 01).

Everything will continue in transformation according to the reasons of HTime and in keeping with our efforts and comprehension. Everything is planned so that the Ego, if it acts in concordance with these reasons, can have an advancement that will be synchronized with the evolution of the Planicio Terráqueo and the Kosmo. Our matter is part of this development; in the future it will be like that which we had before arriving at the Planicio Terráqueo. We'll have the form of a flame. It is in this moment that we will have the konsciousness and the evolution necessary to return to the Kosmo of the Trinak where we will have to complete what was left unfinished."

The sound of a car passing in the street brought her back suddenly to the present. Helena continued walking and entered a small hardware store. The sign posted over the door read "Galvan and Son" in large red letters. She greeted the young man who was behind the counter.

"Good morning, Diego. I came to see if you could help me with a little problem."

"Hi, Helena, how are you? Is there something else that needs fixing over at your place?" responded Diego, smiling. She had chatted with him several times in these months since her return to Remoulins; she had seen something in Diego that reminded her of the same uneasiness that she had experienced as a child, that intuition that says that there's something else beyond everyday life. She had observed that Diego was open to questioning himself and to making changes in order to leave behind the habits that normally dominate us. He had a curiosity for understanding and a sincere interest in the things that she had talked to him about. But lately she had seen an anger that was affecting him, leading him to self-absorption.

Helena explained the problem with the pipes in the cottage to him. He responded in his usual cheerful way, "Would you like me to help you with the installation? I can stop by after I close here." Helena accepted and left to continue with her errands.

The first time that Helena had met Diego was when she had returned to live in her grandfather's house a few months earlier. At first she had stopped by the hardware store to buy the materials that she needed. Diego had always offered to help her with the repairs to the house. He was a youth of 23, with wavy black hair and dark eyes. The features of his face and his swarthy complexion hinted at his Moroccan ancestry. When he was eleven years old he had emigrated from Spain to France with his father, who had gone to work in the vineyards of the region. In the summers Diego had worked alongside him, and had learned much about viniculture and the land. Because he had spent so much time laboring outdoors, he had a strong physique. After six years of toiling in the fields, his father had saved enough money to be able to open the small shop, with the idea that he would be able to rest a bit from his physical labors.

When he had finished high school, Diego had decided to travel with his girlfriend Veronique to the Scandinavian countries. But when the moment of departure had arrived, his father had fallen ill and Diego, being the only child, had acceded to his father's pleas to stay in Remoulins and look after the business. His girlfriend had departed, and with her, many of the dreams he had. Since his father didn't have many possibilities economically, Diego had felt the obligation to care for him as well.

Because he was kind and talkative, people tended to like Diego; nevertheless, he was solitary and didn't have many friends. When he had met Helena he realized that she had the distinctiveness of seeing things very differently than other people. It was something in her that attracted him. On various occasions she had helped him to understand situations in his life and it was becoming customary for him to seek her out when he felt confused, as was recently often the case. He didn't know how to resolve the conflict he felt inside; he didn't want to remain in the hardware store, but the obligation he felt to help his father weighed on him. Many times Diego felt anger and a deep frustration at having lost the opportunity to do what he wanted. Nevertheless, there was a part of him that had taken that decision knowing that it needed to be so. He had been unable to reconcile these two divergent reasons, and it had him in a quandary.

Diego saw that Helena wasn't like the other people he had known throughout his lifetime. She had a total independence; she had neither family nor commitments. Her life was completely dedicated to the studies that she carried out. The only extra that he knew of was her photography. He had sometimes seen her work in magazines; she was an excellent photographer, but she didn't seem to give even that much importance, she just did it as a means of earning enough to be able to dedicate herself to her studies. This intrigued Diego: her ability to follow a path that depended upon no one else, but only upon an instruction that she had tried to explain to him one day.

"I try to fulfill what corresponds to me; it doesn't have to do with my tastes or interests. My inner thoughts are committed to that which manifested me and later projected me in Life. There's a reason for which I'm here and I strive to understand it and complete it."

Her words had left Diego with even more questions in his head. That day, Diego asked her to explain more about this way of behaving; he wanted to see if he could find a way to settle his turbulent mind.

"How can I know what I should do? The truth is that I feel trapped and I don't know where to turn or even what it is that I really want."

"The first step is to accept that there is a greater Voluntad (Will) that governs our development. I've strived for many years to hone a Kommunication with this Voluntad. The key is to accept what falls to us to live in each moment, without imposing what we want, because many times this is contrary to our development. It's that our usual habits always lead us back to what's familiar, even though it's something destructive. There's a fear of doing new things; we've been estranged from our vital reasons for so long and locked into the fantasies that we live in, that we're afraid to leave them behind. We're in a process of evolution in order to one day arrive at a perfection and to undertake this road we need to vanquish fear." Helena observed a certain uneasiness in Diego and continued. "We have an inner guide that can advise us on the decisions that we face, if we ask it and if we pay attention to what it tells us. It's one of the layers that our matter has; it helps us understand what we should do. This guide we call the *Perespíritu, which is a controller of our actions and of our feelings and perception, it can and should relate to us in an effective way so that we comprehend [...] (20 Jan. 09).* Do you remember that opportunity you had to travel?" she had asked. Diego nodded in acknowledgement. "For example, you had the opportunity to do something that you wanted, but you didn't take it. You decided to stay and take care of your father and deal with the hardware store. But you don't know if you did it as a consequence of your habit of making commitments with people or because there was another transcendent reason for your having to be here.

Sometimes the Perespíritu guides us to make decisions that don't seem logical to us, but have a transcendent reason. You should try to understand what it is that you really need to do and recognize what it is that you want deep inside yourself so that you don't feel bad because of it. Right now you're just angry because you're here, he's ill and you've taken charge of him and the store. But your only responsibility is to yourself and to your evolution."

This had left Diego with a great sense of unease.

That afternoon when she had finished her shopping in Avignon, Helena got into her car to drive home. The sky was cloudy and the wind had begun to blow, whistling through the dry leaves of the trees; it looked like it was going to rain later. When she arrived at the house she went upstairs to her bedroom to close the window. She put on a thicker sweater, annoyed by the pain in her back. She stopped for a moment at the desk that sat in front of the large window; watching the dark clouds, she thought of the Genio that looked after the tectonic movements of the Earth, the winds, storms and lightning, and who governs our bodily matter as a whole. She remembered the surprise she had felt when Marion had told her for the first time that these reasons had a konsciousness and were regulated by "workers" that are at the service of the seven Genios, who deliver their energies in the Planicio Terráqueo and in all points in the Kosmo.

She descended the wooden stairs to the living room, observing the path marked by her grandfather over the years. After so many years of use, the wood had worn away where he had stepped so many times. 'It's the same with us,' she thought. 'After traversing the same road so many times, we leave marks; the deeper they are, the harder it is for us to take another route. It's simply easier to follow the path already traced, even though it's dead end. The environment we live in induces us to repeat the same patterns of unconsciousness, forgetful of our reasons. *The Ego is a Real entity projected in an Aparente (Apparent). [...]The appearance in this sphere is so heavy that it confuses the Egos. [...]The Ego must comprehend that this environment is not part of it, because when it involves itself in these voids it becomes confused, it loses its way (7 Jan. 89).* What we need to find is the konsciousness of our acts; without it we keep tracing the same trajectory that we have traversed throughout our incarnations.'

She took some wood from the wicker basket that sat beside the stone fireplace and placing it on the grate, lit a fire to heat the room. She sat on the sofa in front of the window and reflected on how long her journey had been to be able to arrive at this point. She began to play with her ring, turning it round on her finger. The firelight made the green stone sparkle and her memories returned to the moment when she had been walking with Marion through the streets of Paris. One of her favorites had been the Rue de la Paix. There her Maestra had entertained herself browsing in the jewelry stores. She appreciated well-made work and enjoyed looking at the newest designs, the cuts of the gems and the settings. Helena had watched with delight as she observed the pieces; it was like watching a child savoring an ice cream. One day she had stopped to look at a platinum pendant in the form of a teardrop, with a diamond of incredible transparency in the center. The gem shone, giving the impression that it had a spark of fire inside.

"This diamond is extremely old, it's gone through a long process of purification close to fire; you can see it in its clearness," Marion had commented to Helena. They spent a while looking at the other gems while she explained to Helena that inside of that stone was an Ego. Marion had said, *"An Ego introduced into a precious stone is an Ego that has been in extreme heat for centuries and therefore has suffered a great purification"* (19 Sep. 87). Seeing the surprise in Helena's face, she had gone on to explain further,

"I've told you that the Ego is the transcendent part of you, and that our matter is an addition that allows us to live and understand on the Earth. Now you should know that the first matter in which the Ego was projected upon arriving at the Planicio Terráqueo was a mineral matter, that is, stone. The history of the development of the Ego in the Planicio Terráqueo is very long, with beautiful parts and sad parts, and has been forgotten over the course of the centuries."

With this remembrance in her mind, Helena sat at the table to begin writing about the history of man. The hours passed and when she had finished, it was already dark. She capped her pen and closed her notebook; she needed to get ready for Diego's visit.

Outside, the rain fell without ceasing. Helena asked by means of the Kommunication if Diego would arrive. She thought, 'this will be a good opportunity to see if he really has a sincere interest in understanding.' She walked downstairs to the kitchen to have a cup of tea while she waited. A little while later she heard a knock at the door.

"Hello, Diego, come in or you'll end up soaking wet in this rain," remarked Helena, at the same time she opened the door so he could dash in. Diego entered the kitchen and took off his jacket, which was already drenched. "It seems like it's going to keep up like this for a while, I don't think you'll be able to get any work done today. Would you like a cup of tea?" Helena invited him to sit and put the kettle to heat on the stove. Diego extended his jacket over the back of the chair that was next to the stove in order to let it dry and sat at the table. Helena served him his tea and sat across from him. She lit a cigarette, observing how distracted Diego was; he had already added two spoonfuls of sugar to his tea and another to the table. "Now, tell me, how have you been? What's new?" she inquired.

"You know, I feel totally saturated and I've got an urge to go hiking in the countryside for a while. Between my father and the store I'm overwhelmed. I need some air. Don't you ever feel confined in this village?" Diego remained quiet for a moment, pensive, spinning his teacup on the table.

"The confinement is in oneself, Diego."

"You don't know my father. He wants me to stay here with him forever. He even ordered a huge sign with big red letters that says 'Galvan and Son.'"

"Just wait till he puts the lights on it, because you know he's thinking about it." They both laughed.

"The truth is that I'm really sad because he's ill and I don't want to leave him." Diego sighed and, to change the subject, said, "Say, I was reading about the cave they found in Chauvet a few years ago. It's not very far from here and I was thinking of going next weekend to see the area. There are lots of caves there," he said, and more animated, added, "the cave paintings they found are really interesting; I don't know if you've seen pictures of them. It seems that there's a controversy about the exact date of the drawings. They gave them a date of more than 30,000 years but some people say that it's impossible, that humans at that time didn't have the capacity to be able to make them, and that man was barely at the beginning of his development. What do you think, Helena? Have you been to the cave?"

"Yes, I've been there. I went to take pictures a couple of years ago for a magazine. They don't normally let people enter the cave for fear of damaging the drawings. It's an amazing place. You can see in the simplicity of the lines in the figures that the perception of those people was receptive and analytical. The scientists know that they're much older than they say, just like many of the pyramids around the world, but they can't accept it because their theories say otherwise."

"But if they themselves are seeing it, why don't they accept it?" exclaimed Diego, astonished.

"Everything is connected and if they move one piece; even though it may not seem important, the rest are also affected, from the religious ideas all the way to the theories of evolution; in short, the structure of modern society. We think that we've reached the pinnacle of human progress, but there have been civilizations much greater than this. Some realized themselves through technology, others in a spiritual way. If we accept this, then we need to question several things: what happened to them, what did they do, why were they erased from the face of the Earth, what were the consequences of their efforts. And maybe we'd realize that this present humanity isn't as impressive as we think and that the road it's going down right now is leading it to a destruction of its environment and itself. Look, for example, scientists say that we're descended from a branch of apes, but the fact is that the fossils that they've found represent a regression in human evolution. *The mutations that man has suffered in the past confuse scientists, who assure us that man is descended from apes; nevertheless, it is explained to us that it is the other way around, the men whose appearance resembles that of the apes, such as the so-called Neanderthal and others, descend from man[...] (27 Jan. 05).* Our development has been a long process, Diego. Understanding these first humanities isn't totally possible for us, since they didn't have the same perception that we have now, and therefore their interests are foreign to our way of thinking. The first men weren't like we think or how science explains them to us."

"Well, what were they like then?"

"*The first men had a different matter than ours. As he didn't feed himself excessively, but only for delight, his body did not accumulate toxins and therefore was translucent and the colors especially were reflected within him, since his skin was affected by the radiation of*

the sun in a different manner. The solar radiation was also different (27 Sep. 88). In the beginning man had telepathic communication with animals and other men; with the first he coexisted in harmony as he didn't feed on them (he fed himself through respiration and the environment he was in) (1983). The animal was the counselor and friend of man (21 Mar. 83). Men could move about by tele-transportation. Much of the technology that we have now is a poor copy of what we had naturally before, for example, television, the internet, cars, and airplanes.

They were very different from us, they had another type of matter and another type of perception; we can't understand them with the konsciousness that we have now. We simply don't perceive what they perceived. Scientists give the importance to the evolution of our matter, but they should understand that the important part of our sojourn on the Earth is the evolution of konsciousness. Matter is a vehicle through which we can perceive and comprehend; it was given to us to reach a destination, it's not the destination itself. *We should first of all understand that what we observe of man, that is, his matter, is nothing more than a disguise imposed on the Ego, and this disguise changes from incarnation to incarnation to give the Ego a motive of comprehension and analysis, which is part of the instruction that the Ego must have on the Earth, and furthermore, this matter has been perfecting itself gradually throughout the centuries. The first state of being of the Ego was in stone, and then in the animal; first in the smallest, in the insects, until arriving at the largest and from there to the present man"* (27 Jan. 05). *Tal falta el vegetal*

"Don't tell me we were we rocks! How is that possible?"

"Look, if you want to know the truth of our history, first you need to understand why we're here. We're on the Earth to have an education and an adjustment in our interests. We could say that we're like schoolchildren that didn't pass the exam and were failed. We had to be placed in a special site to receive a specific lesson, and as a condition of receiving this education, we had to be limited in matter. And yes, our first matter was mineral."

"Wow. And what do we need to learn in this limitation, Helena?"

"The Ego, through its efforts, has to recognize what it is and what caused it to err in its actions before arriving to the Earth. At that moment it will be able to overcome its current situation and move on to a fuller one. When a child is punished for a prank, the important thing is that he recognizes why he did it, its consequences, and decide not to do it again. It's important for him to take responsibility and not get angry with his parents or throw a tantrum because of the sanction imposed."

Helena passed a plate of cookies, offering them to Diego. He took one immediately, thinking about what could have caused this situation. Helena, sipping her tea, observed him and then said, "Now, tell me, why do you think we have this matter?"

A look of fright crossed Diego's face, and he could only respond, "Hmm, I don't know Helena, I've never thought about it."

"Well, in order to understand this Diego, we need to return to the beginning of our history here on the Earth. The condition of 'sheathing' the Ego in matter is to favor the individual advancement of each Ego, so that it sees and comprehends its own reasons. As a vital part of this condition, each Ego is in a unique apprenticeship. Each should walk alone and without ties and bonds that obstruct this training.

Our state of being in the Kosmo of the Trinak was very different to this; we Egos suffered an enormous change in our structure when we were introduced into the Planicio Terráqueo (Earth) and limited in three dimensions; for that reason we had to pass through a process of 'conditioning' in order to be able to develop ourselves in it. *So that the Egos could survive in the Planicio Terráqueo with a new motive, they were sheathed and were formed in nine bodies [...] (10 Aug. 10).* These nine bodies or casings are what we call our matter. When the Egos were introduced into the Planicio Terráqueo, the layers or casings that the Ego had before were dismembered and it was limited in a dense matter and in three dimensions. As a consequence of this, we don't perceive in the same way as we did before, in other words, our konsciousness was also limited. And our task now is to put these layers of the Ego together again by means of studying the Konocimiento."

Diego was intrigued by what Helena was telling him and he inquired further.

"But what do you mean by coverings of matter, Helena? It's the second time that you've mentioned it. Aren't we made of one single piece?"

"Although it seems that way, it's really not. Our body isn't what we think it is. In reality, it's an *ensemble of equilibriums and reasons prepared for a purpose in a mechanism of control that justifies the assembly of the same (26 May 99)*. It's in a process of constant transformation. It has periods of great changes that follow an order and are subject to the laws that govern our development.

"We can visualize our corporeal form similar to a spheroid or ovoid. Our matter, like that of the Planicio Terráqueo, is not solid or static; *all matter is in a constant movement, the Planicio Terráqueo is in a constant movement in all of its parts and in all of its reasons (28 Jan. 09). The matter is constituted of a limitless multiplicity of triangles that agitate and promote themselves in a spheroid space, in stones as well as in animals and in man (27 Nov. 93).* Plants have another form, since they belong to a different reason than the Egos."

"Why don't we see the Ego and matter as they really are?"

"It's because we don't have the ability or konsciousness yet to be able to perceive the reasons of the Real. In the studies we see that *two things are spoken of; one is Reality and the other is a Fragmented Reality. It is said to us that Reality is when for example an object is complete because within the Fusion it has passed through all of the necessary steps of evolution within the trajectory that it should take and it is seen as a whole.*

It is said that a Fragmented Reality is when an object or other reason has not completed all of the steps or cycles of evolution through which it must pass, that is, it is still in an intermediate point of its training or evolution and therefore it is said that it has not arrived at its fulfillment or its Reality. It is said therefore that this is a Fragmented Reality and we are given the example of a complete apple and an incomplete one (9 Apr. 08). The Fragmented Reality is the illusion or the fantasy which we perceive and accept as our world. *Our corporeal form is in part fictitious. […] Our vital form is like an egg from which issue forth ranges of lux (light) when it has arrived at a maximum […](13 Jan. 09).*

"Well then, the spheroid or ovoid of our matter is comprised of nine layers or casings, as if it were an onion. The matter of each layer has different degrees of density; some are formed of dense matter that we see and touch. Others are of a more subtle matter and therefore they have been named 'spirits.' They're nourished by different reasons. *The physical matter needs to eat the matter that we normally consume, but the spirit bodies are in another dimension and need to eat the nourishment that the planets furnish in the entirety of their emanation (Oct. 09).* Each layer has its own sonorousness, function, time, and konsciousness, and they are in different dimensions. Nevertheless, not all of the layers are presently complete; they will evolve in accordance with the HTimes of the Planicio Terráqueo. Because of this, our matter will have another form at the end of the Seven HTimes. These layers are interconnected and form something like a 'sieve' that surrounds the Ego. This sieve is permeable; it allows information from the environment to penetrate and lets out that which the Ego projects: thoughts, words, emotions. In rocks, it's more 'closed' than in animal matter and this in turn, more than the human. There are many reasons in the environment that we don't perceive; they're all there, but we don't have the ability to be able to be cognizant of them. For this reason it is said that we don't invent anything, we just take in that which is already projected."

60

"And what are the layers of our matter like? What do they do?"

"Imagine the mechanism of a watch that has many gears of varying sizes, but they need to be synchronized with each other to be able to function. They all have a reason for being. Well, it's like that with our matter on the Earth. The nine layers that comprise it are like the gears of the watch that should be in harmony and synchrony. The Ego is the layer that should synchronize them all, but most of the time this isn't the case. There's a study that explains that the greatest imperative of the matter *is to understand and to give itself in submission to the Ego. The konsciousness of the Ego is distinct from the konsciousness of the matter. Its 'orders' are different, but within the dimension man, the reason of the Ego should prevail, however it is not so; matter has prevailed over the Ego. The konsciousness of the Ego is what ye call spirituality and the konsciousness of the matter is materiality. The*

Perespíritu is the mediator between these reasons, which are more in the Ego and less in the matter (17 Sep. 99)."

Upon seeing the confusion in Diego's face, she asked, "The Ego is formed primarily of Lux Sonora (Sonorous Light), right?" Diego nodded in agreement. *"The Ego is a sonorous entity; the matter is adapted to our sonority. The matter receives the sonorous impacts of other people. The world in which we live is sonorous (23 Aug. 90). Matter is like an acoustic receptacle; this acts like a mirror, gathering and reflecting indefinitely the sonority of the Ego (17 Nov. 99).* The interaction of the layers of our matter helps the Ego to comprehend and assimilate the information that it gathers from its own projection and from the environment. Remember Diego that everything we talk about concerning the Ego and the matter has to do with sound.

"We begin our lesson with the Ego itself, what we call **Yo Ego Deidad**. The Ego is what we are, *it is the transcendent particle (15 Apr. 07).* The 'Yo' (I) is what permits that you recognize yourself, for example, 'I am.' *It recognizes that it is, and is the beginning (15 Apr. 07). The Yo is the comprehension of perceiving oneself, and within all of the reasons situate itself as a preferential point, that is the beginning of konsciousness[...] (26 Oct. 07).* The part called 'deidad' has to do with our vital function. The word 'deidad' refers to a vital makeup that the Ego has: the action of giving and receiving, of projecting itself and contracting itself." She drew the figure of a spiral of attraction and another of expulsion in her notebook to demonstrate to Diego the centripetal and centrifugal movements. "With the external movement, the Ego projects itself in order to grasp its surroundings. With the internal movement, the Ego gathers back into itself to understand its condition."

Helena decided to take a break. She stood, saying, "You know, I'm feeling a little bit chilly. Why don't we go into the living room?" They eased into the two armchairs that were situated in front of the window. She placed the plate of cookies on the coffee table.

"These cookies are delicious; they're from Marie-Claude's bakery, right? Only she can bake like this," commented Diego. Helena smiled as he lifted another cookie to his mouth. They remained for a while listening to the sound of the falling rain.

"And what are the other layers, Helena?" Diego was intrigued by the talk, so much so that he had forgotten all about his frustrations, although not about the cookies and he served himself another one as he waited for Helena's answer.

"We'll begin with the layer that carries the record of all of our actions; it's called the **Espíritu Huella**. *It is this which predicts the presentation of man and within it the Supreme Entities act in harmony (7 Feb. 01)*. In this layer is recorded all that we have lived since our first home, the Kosmo of the Trinak. It has the *konsciousness of memory—in the present it is of this incarnation, but in the future, it will be of all of the incarnations (28 Feb. 05)*. It's as if it were a movie in which all of our acts and thoughts are recorded throughout our incarnations. When we die or Trebolar, this part endures with all of the information to manifest itself once again when we enter into a new matter. In each incarnation the information guarded in it manifests itself in us and it's seen in some of our physical characteristics, in our habits, and in the personality that we've formed through the centuries. *It has a latent similitude of all of the speculations of the centuries; because of some problem or motive, these reasons of yesteryear may surface in the present (9 Aug. 90)*. The Espíritu Huella manifests itself in dreams, in our thoughts, and in memories that many times are awakened by simple things, such as hearing a piece of music, seeing an object, or smelling an aroma. It attracts these memories and it presents them to us. Many times we don't understand them and they pass us by. Right now they manifest themselves without an order, but in keeping with the advance of the Ego in its development, it's possible for us to have the sequence of all of our experiences present. *The memories will be full in us and will form a multiplicity. We will have omnipresence (Nov. 10)*.

"The **Perespíritu** is our inner guide of which I spoke to you earlier. This layer is a very special reason because it's not part of the Ego itself,

62

or of the matter, rather it's a 'Kosmic entity' which was delivered to us to advise and guide us in our actions here on Earth. It has its own process of development and evolution and upon fulfilling its service, returns to the reasons to which it belongs. It regulates the Kosmic reasons that arrive to the Ego at specific times and the manner in which they manifest themselves in the body. Through it, many reasons are projected to our matter, from chemical and hormonal to reasons for the development of the Ego. It knows to where we should direct ourselves, it guides us through the path of karma so that we fulfill it. You should ask it about what you need to do.

The Perespíritu is not the same that we have always had; it changes from one incarnation to another. Sometimes it does not accompany a matter from birth to death, it is changed for some reason specific to the evolution of the Ego, which receives the assonance of a new Perespíritu; this manifests itself in radical changes in the life of the pequeño (for example, in the case of Paul Gauguin), who had a drastic change in his life *(9 Aug. 90). [...] It should manifest itself as a protector, a guide towards the new paths, and should communicate with us constantly (15 Jul. 95).* It's vital that man attunes himself to this guide. You can request that it attracts the reasons that will serve for your development, Diego. It communicates with us by way of nervous impulses that translate into sensations, emotions, and thoughts. The Perespíritu can guide your dreams, attracting memories and images that aid our comprehension.

"The layer **Materia (Matter)** *is the Kosmic matter or Kosmic dust of which the matter is formed (Carpeta Dorada). The matter is a vehicle of interrelation between the Ego and its excellences and the inherent reasons of the Planicio Terráqueo (17 Aug. 90). It should be subject to the actions and considerations of the eight bodies, with a submission of complete surrender to the same (15 Jul. 95).*

"The **Yo Interno** *are the organs that establish a function from the brain to the kidneys, liver, spleen, pancreas, etcetera (23 Aug. 90). The vital organs act by means of their own criteria, but are identified between one another. Each organ has an intelligence of itself, the*

63

liver is the liver itself, it forms a point and has an awareness of itself, an awareness of the immediate environment and an awareness of the environment as a whole (28 Aug. 90).

The brain of man registers the whole of the actions of the bodies of the Ego. [...]The brain is like a magnet, it collects all that surrounds it in the environment, it perceives all that happens around the pequeño, although it does not discern it (28 Feb. 05). Each organ has a konsciousness of itself and of the others. Because of this, when a surgeon extracts one organ, it affects the entire aggregate of matter.

"Then there is another subtle layer that's called the **Yo Konciente**. *It is a filter that is reflecting what the Ego senses in the Planicio Terráqueo (13 Aug. 90).* It's located in our heart and regulates the pressure of the circulatory system. In this layer *are manifested the sensations, emotions and feelings. Each of these actions mobilizes the states of konsciousness (15 Jul. 95).* However, it presently gives us a distorted assessment, since *now the pequeño perceives more through his feelings and emotions, and not by way of a Kosmic reasoning,* as it should *(11 Oct. 95).* The Yo Konciente has many levels of perception and translation that are reflected in the osseous, muscular, nervous, hormonal, and cerebral systems. It has a relation with what the matter experiences in the environment, and translates these data so that the Ego understands them. This action occurs by means of the brain. *We can say that the brain collects the sonority of the Ego, the sonority of the matter, in which enters its own sonority, and that of the Yo Konciente (5 Oct. 93).*

The Yo Konciente perceives through what thou callest senses, and this is one level of perception. When it perceives its projection it manifests it in the cerebral organs, and it is by means of the continuum in what thou callest relation and nervous confluence.

The Yo Konciente has an imperante* (*ruling laws), that of manifesting itself towards a maximum reason, which is to understand. [...]The perception of feeling is an ephemeral reason that is connected to a very elevated reason. In the future, the Yo Konciente will develop more levels of perception, and this will fortify it and enlarge it (8 Jan. 96).

"The **Espíritu Carne** is located in the blood and in *all that contains tissues, nerves, tendons, and reasons of the skeleton (23 Aug. 90). It is the conjunction of all of the matter (Carpeta Dorada).* It's highly sonorous; in fact, the red blood cells are the bearers of sonority.

"The **Espíritu Materia** *is the unification of all of the matter with all of its perceptions (28 Aug. 90). It is the aggregate of all of the matter in plenitude (Carpeta Dorada).*

"Another layer is the **Espíritu Psíquico**. In the Kosmo of the Trinak *the Yo Ego Deidad and the Espíritu Psíquico were one (Carpeta Dorada).* This layer functioned as the sonorous receptacle for the Ego; it needed it to be able to understand itself. But this relationship changed when we were introduced here. Now the Ego understands by means of the whole of its matter. *Before the Ego had matter it projected itself in the Espíritu Psíquico which was like a condenser and transmitter of the energy of the Ego. It was the receptacle that gave force and power to the Ego like the matter does now.*

The brain is an imitation, a substitute for that sonorous receptacle that was in the past (11 Jun. 88). In the Espíritu Psíquico is stored all of the evolution that the Ego has had; it is not erased, it simply remains in a place that for now is untouchable. It is 'separated' from the Ego, although it is still present."

"Wow, Helena, it's amazing that we don't even know what our body is made of," exclaimed Diego, astonished by the chat. He thought for a while, reflecting on what Helena had said. He realized how much he had left to understand about the world and himself. A question came to his mind and he asked, "Then, why isn't there en equilibrium between the matter and the Ego, if it's all designed that way?"

"When the matter imposes its needs upon the Ego, the Ego enters into a great imbalance. *The Ego is diminished when it subordinates itself to the reasons of the matter (17 Aug. 90). The matter perceives emotions, sensations, etcetera, but it does not comprehend the real Kosmic changes or assertions of the Ego. The Ego is always acting according to the modifications that are vital for its condition of being (23 Jul. 99).*

The brain should be a recipient that will perceive the nine bodies in plenitude. [...] Intelligence is the interrelation of the brain with the Yo Ego Deidad, but the brain does not know the Ego, it only receives data; it has a perception and an intelligence. Within the reasons of the brain is the misalignment of the past and the future.

When the Ego is not connected with the brain, the matter grows weary, because the Ego 'goes to sleep' (and separates from its matter), *and this loses its vital energy. In this case the matter imposes itself and looks for sensations that it is accustomed to" (9 Aug. 90).*

"And when the Ego doesn't understand, the matter becomes maladjusted?"

"Yes, Diego, the Ego should always act according to the reasons that are favorable to its evolution, but it doesn't do it because it is unaware of them or because it continues in rebellion. Look, I'll give you an example. The Planicio Terráqueo was formed so that we could comprehend and advance. But we no longer understand it or how we're linked to it. There are many interests that want to destroy it by contaminating it, maltreating it, thinking they can do what they want with it. *The Ego, reigning in its attitude of self-esteem, does not understand its surroundings or accept the reasons that propel it toward the reason created for its understanding (11 Jan. 01).*

What we don't understand is that in damaging the environment, we do ourselves damage, since the true structure of our matter has an inveterate relationship with the formation of the Planicio Terráqueo. It's not simply a question of damaging our home; it's a question of provoking an immeasurable harm to ourselves.

The formation of what thou callest virus, insect, animal, whether it be aquatic or terrestrial, and therefore, man, is in its complex structure the result of an earlier manifestation. [...] This realization is the Earth itself. We tell thee that as a fish has gills, and man has liver, spleen, etcetera, likewise, the Planicio Terráqueo is configured in similitude. The structure of what is called the Earth resolves in itself each and every one of the wonders that thou knowest (21 Apr. 98).

Because of this relationship, this is the only place in which the Ego Terrual (Earth-constrained Ego) can develop, although mankind wants to escape from this Planicio Terráqueo to go in search of a new home in space. His material structure, the reasons of his transformation within it, and his konsciousness are an integrated part of the Earth. It's not permitted to leave the Planicio Terráqueo until we have completed our development within the Seven HTimes."

Diego remained silent, contemplating all that this implied and the lack of consciousness with which we live. He looked up and saw the sketch of the spiral in Helena's notebook and remembered something that she had told him earlier.

Outside, the rain had ceased and the odor of wet earth entered through the window. Helena knew that this was the moment to finish her talk; Diego already had too much information to assimilate. She stood up to put more wood on the fire. Diego helped her arrange it and noticed that a stone in the wall of the fireplace was about to fall. He told Helena that it would have to be reset and the rest would need to be checked so that they didn't all come loose.

"If you like, I can take care of it tomorrow too. I can bring the material to fix it." He checked his watch and saw how much time had passed. "It's so late! How could the time have passed so quickly? I had no idea!" He had forgotten his problems and worries while talking to Helena. He'd also completely forgotten about his father. Feeling pressured, he said goodbye to her, thanking her for what she had taught him. He said, "I'd like to be able to talk to you some more about this, if you'll let me Helena. But for now, I'll be back tomorrow to take care of the installation. Is the same time ok?"

"That's fine, Diego, I'll expect you tomorrow." They shook hands and she walked him to the door. She stood in the entry while he started the car and headed out to the road. She reflected on the resentment that Diego felt towards the things that had fallen to him to live. He hadn't taken the time to analyze them in order to comprehend them. In spite of the emotions that were befuddling him now, she saw that Diego was open to accepting direction, to making changes that would help him to find a new way to see himself in the world.

Helena closed the door and walked through the kitchen to the room that her grandfather had used as a storeroom, but which she had converted into a darkroom. She had to develop the photos that she had taken in Arles in order to send them to the magazine. She enjoyed working at night as she was able to concentrate well. And she still had to study about Diego and consider if it was a possibility for him to enter the studies. She thought that maybe his opportunity was coming, and remembered how hers had arrived all those years ago.

Chapter 4
From Helena's Book:
Where do we come from?

Thinking about the talk she had had with Diego, Helena sat at her desk to continue writing the book. She remembered the questions she had asked her Maestra all those years ago and began to jot down in her notebook:

> We continue with the questions that men have asked throughout the ages. Where do we come from? Why are we on the Earth? What is the reason for the experiences that we have here?

To understand this, we return to that distant time when we still lived in harmony with our reasons and with the Konceptos (Koncepts) that direct everything in the Fusion.

The Egos before coming to the Earth, in the Kosmo of the Trinak, were sonorous builders and sonorous modifiers (21 Oct. 88). We projected our thoughts in the form of sounds in order to comprehend ourselves and our environment, by means of the actions that we recreated.

There, *the Egos were in different levels of awakening. There were incipient reasons, the unborn (which is insinuating that it is going to be born and in a period of gestation) and the already born (children, youngsters and elders, the most awake) (5 Aug. 89).* All of us were in a distinct process of development, each one fulfilling the function which corresponded to us.

Since we had different functions, we were grouped in Huestes (Hosts) according to our purpose of realization, as in the present there are different functions among men: architects, bricklayers, artists, etcetera. In each Hueste there were distinct levels of development and konsciousness, and we were accommodated in a hierarchical structure according to that development. In the Huestes there were more evolved Egos that had a greater capacity and consequently, the responsibility to help the others comprehend and advance. In our manner of understanding, we could say that they were the leaders of the other Egos.

Nevertheless, all of the Egos were beginners in the reasons of Life. In order to be helped and guided in our development in the Kosmo of the Trinak, we had a "tutor," who was an entity of greater investiture.

If we were in harmony with our reasons, why did everything change? Every particle that is within Life is learning; in the same way this tutor was in a process of evolution. There came a moment in which he, confusing his actions, appropriated for himself reasons that did not correspond to him, and, greedy for power, rebelled against his programmed mission. He gathered the Egos together so that they would follow him in his rebelliousness. We received a warning from the Konceptos to correct our actions, but we didn't listen.

In his arrogance, he and the majority of the leaders of the Huestes used their power and led the other Egos to an error. We could say that the Egos allied themselves and formed structures that shouldn't have been. This action altered the function of the Kosmo. We entered into an erroneous endeavor, believing that we could dissociate ourselves from the Voluntad (Will) and "create" for ourselves. We took the graces that we had and used them to "imitate" something that was beyond our function. We erred in thinking that what we built was Real, and that we could overstep the limits of our preparation.

When we say Real, we can understand it, in simple terms, as the Reason of the Beginning in which is "contained" the formation and possibilities that a particle has, and the Aparente (Apparent) as the action of that particle in which it experiences an evolution and transformation. As an example of the Real: *we can describe it as a rocket that is the beginning and when it is lit it establishes manifestations as in the case of fireworks, of colored lights; that which we see and perceive is part of those fireworks, but they are not the "rocket" itself, and they last an instant or some minutes, to later disappear and transform themselves, since the manifestations of the Vital are continuous[...] (4 Jun. 09).* We should understand that all that we do is within the Apparent, and we do not do or touch anything in the Real. *The Universes, the Galaxies, that surround us are not Real. They are in Appearance; nevertheless, something remains so that they pass on to be Real (Carpeta Dorada). The Egos are Apparent Deidades in preparation to become Real Deidades (Carpeta Dorada).* (Deidad refers to the action of giving and receiving).

When the Egos entered into disharmony *[...] it was a chaotic epoch in which all of the Huestes in great part accepted that rebelliousness and all of this ended in a trial* and in the separation from their tutor *(10 Aug. 10). In the Kosmo of the Trinak, when we created this error, something was born in us that was not congruent, since it had not been in us, but it happened: Fear,* which resulted from knowing that a Law had been broken *(1 Apr. 89).* So that we would not continue altering the Kosmo, the Egos were "subtracted" from the Kosmo of the Trinak, isolated and limited. "Subtracted" *[...]means "they are separated from their original environment," to be submitted to a greater or subtractive consideration. The subtraction is an*

apparent limitation, to measure and accelerate the process of activation (1 Mar. 94).

With this subtraction we lost the Kommunication with the Konceptos and we entered into a period of isolation. During this period, the Earth or Planicio Terráqueo was planned and formed.

The Earth was conceived as a special school, so that the Egos would come to realize our error and we would find the correct relationship inside of ourselves and with the Supreme Entities again. When it was formed, *we were introduced into the Planicio Terráqueo to separate us, as we were no longer in harmony (Carpeta Dorada).*

The Huestes were separated and each Ego was encapsulated within matter so that it would have a unique and specific adjustment and apprenticeship. The process of finding this adjustment is what we know now as our life on the Earth, which started thousands of millions of centuries ago.

The Ego is a particle that did not have an exact accommodation, there was a maladjustment. In the moment of entering the Planicio Terráqueo, it was dispersed and its particles were moved as part of a lesson that consists of achieving once again an adjustment of these particles to find its power again in order to acquire all of the potential that it had before. This apprenticeship is the surmounting of the Ego. It was dispersed so that we put it together again and that saturation will be useful so that it does not err again. Upon recognizing the value of this overcoming because of the effort involved, the Ego will use it for its benefit (29 Nov. 89). As part of the process of entering the Planicio Terráqueo, the structure of the Ego was modified; its layers or coverings were disarticulated, and the Ego was encapsulated in matter that was three dimensional and denser than that which it had previously.

The first matter in which the Ego was projected on the Earth was rock. So began the apprenticeship of the Egos in the First HTime. *The awakening to life within the earthly reasons was when the Ego was manifested in stone (10 Jan. 01).*

Upon the Ego being introduced in matter, a state of conjoining with the same was established, so that it would function as an act of limitation. Thou shouldst understand that this limitation was much greater in stone, a little less in the animal, and somewhat less in human matter (2 Jun. 00). Part of this material condition is to propitiate the individual advance of each Ego. As part of its lessons it would have to pass through fear, *so that the Ego would respond when it doesn't understand, and as a result, intuit that not everything was comprehensible for itself (9 May 96).* This fear would help it find humility again.

In stone, the konsciousness of the Ego only allowed it to perceive the past, so that it would have an introspection and understand what it was that led it to the rebellion in the Kosmo of the Trinak. *When the Ego is projected in stone, it has a global reason of all that has occurred to it in the Planicio Terráqueo, that is, it knows why it has been fused in stone (4 Dec. 82).*

When the Ego finally understands and accepts detaching itself from the stone, it has a sensation similar to that which the pequeño (little one) has in the animal and in man at the moment of its birth, suffering as a result a sensation of displeasure, amazement, or pain (10 Jan. 01).

When the Ego has understood and its konsciousness permits it, the next step in its development is that of projecting itself to animal matter. This stage is comprised of many experiences until the Ego arrives at comprehending its state of being.

73

When it passes into the animal, it promotes the equilibrium of itself and of its own movement and comparison, fortifying the need to connect itself to the environment (4 Dec. 82).

Animals have another level of konsciousness and different lessons than those that are had within stone; they have an understanding of their surroundings and the perception of the present, but not of the future.

Upon arriving at having the konsciousness of the "I", the Ego projects itself from the animal to human matter. In the human form it can have the comprehension of its experiences in stone and of those in the animal, unconsciously, and thus it can ask about the reasons of its past.

When the first men in those primordial ages were qualified and sufficiently developed, they received the teachings that were delivered in the First HTime, so that they could begin upon their journey of comprehension and evolution.

Nevertheless, even now we have not accepted the condition of being in the Planicio Terráqueo or the teachings that were delivered in the First and Second HTimes. Since our arrival, some Egos have rejected the conditions of the apprenticeship in matter and in the Planicio Terráqueo. Man has continued conspiring in rebellion, destroying his environment, and making his experience on the Earth ever more problematic.

When we speak of the different HTimes, it's because the Planicio Terráqueo was conceived by a great Koncepto so that it would develop itself in Seven HTimes. *In the same way that an architect upon fabricating a house, knows approximately what will be its duration, thus the Omnipotente Padre Eterno Jehovah (Omnipotent Eternal Father Jehovah) upon designing the Planicio Terráqueo*

knew how much "time" it would last (Carpeta Dorada).
HTime is a Sonorous Akt and because of this we may say
that each HTime or tercia is a melody, a tone, a sound, a
level of konsciousness, an awakening in the evolution of
the pequeño (18 Oct. 91). The development of the Planicio
Terráqueo and all of its reasons are already formed within
these HTimes and they will project themselves in the
course of the centuries. It's like a book already written and
upon turning the page, the new reasons make themselves
"visible," manifesting themselves in the changes in the Earth
itself, in the plant kingdom, in the animals, in our matter,
and principally in our konsciousness. Our konsciousness
will transform itself during these Seven HTimes, giving
way to the possibility of perceiving reasons that now
are beyond our level of evolution. All of the lessons and
variations that the Egos are going to have in each HTime
are already programmed, but the manner in which they
present themselves depends upon the disposition and
acceptance in which humanity finds itself. If we understand
and complete the lessons of each HTime, these will be
short; on the other hand, each lesson not completed makes
them longer. This is why the First HTime lasted, according
to our understanding, millions of years.

In the same way as the Planicio Terráqueo, matter is
programmed to experience a change and development that
will promote the understanding and evolution of the Ego
in each HTime. The matter that we will have in the future
won't be like this which we know now. And as it seems
difficult for us to remember what our first matter was like,
we can't imagine how it will be in the following HTimes,
although it will resemble the ethereal and graceful matter
that we had in the past.

Helena paused, thinking of how the Planicio Terráqueo was a
sphere of repercussion in which each one of man's thoughts and acts
experiences a resonance and multiplication so that it can be seen and

analyzed in the reflection. 'What will the changes coming in these next years be like? Will we be able to adjust ourselves in harmony to the transformations in the Earth, matter, and the Kosmo?'

Chapter 5
On the Way

Our leaders have decided to spill all the oil possible in the seas
so that the ladies have the complexion of schoolgirls.
RECIPES FOR HEALTH

The hands of the clock which hung over the counter turned slowly
and time seemed to drag itself out for Diego. He thought of
the unfinished tasks that were waiting for him at Helena's house and,
impatient to see her, he decided to close the hardware store early.
He'd need to take mortar to fix the fireplace and his tools to install
the pipes on the patio; he wanted to get started while there was still
a little light. But what he most longed for was to talk to Helena. He
had spent the day thinking of the things that he wanted to tell her
about the goings-on in his life. He asked himself, 'what's the purpose
behind everything that's been happening to me?' Analyzing it in this
moment, he realized that he had made many decisions only because
he felt an obligation, and he had never really thought about the why
of things.

In the last few months Diego considered that his life had unfolded
itself via unexpected circumstances. The changes he experienced had
been abrupt, causing a great agitation in him. The first years of his
life had been relatively tranquil and without great problems; he lived
content in his home in Navarra with his parents, going to school in
the day and playing with his friends in the afternoons. His childhood
had been happy, to his point of view and, although he hadn't grown up
with many belongings or luxuries, neither had there been privations
or great suffering. Nonetheless, everything changed suddenly when
Diego was 13 and his mother had fallen ill. In just a few months his

father Esteban and he had been left alone when she died. His father had fallen into debt with the hospital expenses and he had to sell his small farm. Then Esteban had taken the decision to change his life and look for a new home, trying to distance himself from the pain that he felt. A friend offered him work in a vineyard in France and he accepted without asking Diego. They packed what was necessary, sold the rest and went to live in France.

When Diego arrived at his new home, he had felt uprooted and without support. He didn't speak French and it had cost him a lot of effort to be able to get along with the other children in the school. The other students had ignored him or treated him with disdain; for them he was just a poor immigrant that they didn't understand. Diego hadn't really grasped what his schoolteachers were teaching him; he felt stupid and alone. He missed his home, his friends, the animals he had raised and above all the landscape that he used to roam, scaling the rocks and fishing in the river. He began to feel anger, as much for the death of his mother as for his father's decision to take him to a place where he didn't know anyone and he couldn't even communicate. The two events had seemed unfair to him. He forced himself to study and quickly learned the language, and with this he was able to make friends, attracting them with his good humor and his willingness to help. In the summers, when he had labored alongside his father in the vineyard, he learned much about the region. When he had some free time he disappeared for hours walking alone in the countryside. In this way he developed a deep affection for the area's nature, for its red and ochre soils, the impressive limestone rock formations and the vegetation. His journeys through the region to see the ruins and the old buildings had awakened in Diego the desire to travel the world.

Diego yearned for those moments of freedom that he had felt in the past. The impossibility of taking the trip that he had so much wanted, the confinement in the business and on top of it all, caring for his father, had him disgruntled and filled with a malaise that he couldn't shake off. He felt trapped in a life that he didn't want, detained by some force that he didn't understand. He remembered what his mother had told him, "If you treat people well and you do the right thing, looking out for their well-being first instead of your own, life will reward you."

'Well then, I don't understand why everything turns out so badly for me. Is it just bad luck? I've tried to be good,' he thought. He didn't understand it and in time the frustration that remained inside him had caused him to be discontented and resentful. Now, he just wanted to know why he had to live all of this.

Diego wanted to talk to Helena about it. Surely she would be able to give him some precise words of advice about what he should do. He was amazed at how he could remember the earlier conversations he had had with her. They were recorded inside him, but he could only remember them when his emotions didn't cloud his mind. Last night Diego asked many questions of Helena but he didn't dare to really express his deepest concerns. 'I feel embarrassed talking to her about the things that I have inside.' But then he realized that it wasn't timidity that stopped him, rather it was a fear of confronting his life. He sighed and thought, 'maybe today I can do it. I'm sure that Helena can help me.' With this thought in mind, he called her to ask if he might stop by her house earlier than planned.

A little after 4 p.m., Helena heard a car coming up the drive to the house. She opened the door and saw Diego getting out of his old Renault. Helena greeted him and invited him into the kitchen.

Curious to know if the repairs that he had made in the roof had held up, Diego asked her, "That was quite a downpour last night, wasn't it? Did you have any leaks in the house?"

"The sound of the rain helps the plants grow. Isn't that amazing?" Diego just looked at her, unsure if she had understood the question. "It seems as though everything is fine, Diego. Well, I'll leave you to your work; I need to make a phone call." He watched her go, unable to say anything. He felt uneasy and he thought, 'it seems like my questions are going to have to wait for another day.'

Helena sat at her desk and reflected briefly upon the agitation that she had noted in Diego. She knew that he was opening himself to the idea that the world, just like his life, wasn't what he had always considered it to be. But he would need to get to the point in which his desire to understand was stronger than his desire to fit into human reasons. He was in the midst of a process that could lead him to make a transcendent decision. She decided to leave him alone with his uneasiness for the time being. Diego would have to see his imbalance and insecurity for himself if he wanted to understand what he really desired.

She uncapped her pen to continue with the book she was writing. Opening the notebooks of studies, she found one that spoke of the transformations that there have been in the Planicio Terráqueo and in the konsciousness of man.

'Your Planicio Terráqueo has suffered in its structure eighteen transformations, all of them emphatic and which have marked in the same, unsuspected changes for thee. That which thou callest rotation has suffered intense variants, sometimes its day was slower and others faster, but it is said to thee that this is relative to thy current capacity of comprehension.

Thou must know that each change suffered by the Planicio Terráqueo, provoked in the Ego a distinct accommodation of orientation in that which thou callest konsciousness. The perception of earlier humanities was distinct to that which thou hast in the present, as their reflections and resonances were not the same as humanity currently has. Approximately sixteen-thousand years ago, a reference is made to the present humanity, in which was manifested a konsciousness in development under a perception which forms part of an angle of the same konsciousness. Thou must know that six times has the rotation of the Planicio Terráqueo been in one or another direction, and its polarity has been inverted on two occasions. In the following years, when it loses its current polarity, a new konsciousness will be manifested, since this perception of instruction was not understood. It is in these changes, when the atmosphere of the Planicio Terráqueo grows or declines' (27 Mar. 98).

Diego felt frustration when he saw Helena go upstairs. He decided to concentrate on his work and went outside to fix the pipes on the patio. It took him longer than expected; there were problems at every step. He entered the house and began to fix the fireplace. He checked each one of the stones and saw that three more were on the verge of coming loose. He had to remove them, clean the wall and affix them again, but the aggravation that he felt inside wouldn't allow him to focus. Then, a rock fell to the floor, breaking; on top of it all, he had stained the floor with the mortar. 'Damn it! I wanted to do a good job for Helena and everything is going wrong! What's happening to me? What the hell!' Nonplussed, he went out into the garden to take a breath of air and sat on a step of the patio to clear his mind.

Night had fallen and he sat looking at the stars that filled the sky. The moon was almost full and a cold silvery light bathed the countryside. Diego felt a profound loneliness and in that moment noted the weight of the anguish and disillusionment of the last few months. He was surprised to hear himself asking the moon for help and clarity to be able to find his way. 'Wow, what am I doing talking to the moon?' he thought. But at the same time it seemed to be something natural to him. 'What's happening to me today?' In this instant he heard Helena's voice calling to him from the kitchen.

"How's it going, Diego? Did you finish?" she asked, leaning out of the back door of the house.

Diego turned and saw Helena standing in the doorway. The moonlight fell over her and he saw that the stone in her ring gleamed. Suddenly he perceived her differently, with another appearance, dressed in a long tunic. She gave him the impression that she was a woman of authority and with a great responsibility. Surprised at what he was seeing, he rapidly pulled himself together, shaking his head and knocked a flowerpot to the floor, which luckily didn't break.

Nervous, he responded, "Everything's fine, Helena, I'm just finishing." Diego gathered his things and went into the kitchen to wash his hands. He entered the living room where Helena was already checking the fireplace.

"It's a good thing it didn't fall on your head." She let out a laugh.

Diego blushed and said, "It's been a strange day. I just wanted everything to turn out well, but it's been complicated."

He remained standing in front of the fireplace, waiting for Helena to say something, but she just knelt and started to put some wood on the grate to light a fire, while saying, "In spite of all of the difficulties you had, everything turned out ok. Thanks for coming."

Diego didn't know what to say. 'Is she telling me to go? But I haven't had a chance to ask her anything.' He was disconcerted and preferred not to say anything else. But without thinking, he heard himself saying to Helena, "You know, I'm thinking of going to the caves near Chauvet this Sunday. Would you like to go?" He was shocked to hear the words come out of his mouth.

Helena turned to look at him, appreciating his effort. She responded in a soft voice, "Yes, I think so; we can walk for a while. It's an interesting area. Is eight o'clock good for you?"

Feeling a huge sense of relief, Diego answered, "Yes, that's fine Helena. And I'll bring some baguettes for the road. I'd better say goodbye now before I screw up again and I'll see you Sunday." They both laughed.

Helena walked him to the door. Diego wanted to apologize once more for the problems, but before he could speak; Helena said, "And wipe that look off your face, don't worry. Everything starts to arrange itself when we know what we want; it's not knowing that disturbs us. The key to everything is to understand what we're looking for. Look, there are many small reasons that hinder us and drain our energy.

Normally we spend our life in these insignificant reasons and we don't see that there is something beyond them. We should arrive at a profound reason, which is our destiny. *When that reason is awakened, the Ego leaves everything and does that. Destiny is what we set and preset throughout the ages; nevertheless, there exist adjacent ideas that distract us from that reason and hence we detach ourselves from the vital reason"* (8 Apr. 88).

Diego listened to her words, attentive. Once more Helena had grasped his disquiet and he felt exposed before her. Out of the blue he muttered something about his father who was waiting for him and left quickly, forgetting his toolbox on the floor.

When she closed the door behind Diego, Helena saw the box that he had forgotten and remembered how she herself had felt off-balance and distressed by the changes that she had faced when she had begun to talk with Marion. Helena remembered a study that Marion had pointed out to her. *"Disenchantment is the manner of feeling uncomfortable with oneself. Distress is the imbalance of matter. [...]We were conceived with a certain purpose [...]. Because*

of this we have inside of us an order that must be fulfilled. That is what manifests itself as a distress in man. The angst that man feels at living is because he is not being useful for the purpose for which he was created [...]. We will find true satisfaction in what gave us origin. The means of satisfaction is evolution" (2 Jul. 88).

When Helena walked downstairs Sunday at daybreak, the ashes that remained in the fireplace languished. She walked cross the flagstone floor to the back door and peered out into the garden. When she opened it, she saw that the chrysanthemums were glazed with a thin layer of frost. The first rays of the sun still hadn't begun to warm the small stone house and she felt the cold of the dawn on her face. She bundled up and stepped out onto the patio.

The pale tones of rose, coral and lavender began to illuminate the clouds that floated over the garden. Helena observed how everything slowly revealed itself with the arrival of the light and remembered something that her Maestra Marion had explained when she talked about the process of painting. "It begins with the darkness in the background of the canvas. The figure is there but it still hasn't defined itself; it's veiled. Little by little the painter removes the shadows and puts in the highlights that define the form. It's the Lux (Light) that defines the form." Sometime later she had understood this as a metaphor that described the process of the person who approaches the Lux. She reflected, 'the Lux is what defines us, it gives us movement and purpose; it is our beginning. The Konocimiento grants us the possibility of understanding our true reasons; it is the tool that helps us remove what has maintained us in darkness for so long.'

She wandered around the small garden and began to emanate the plants, letting the energy flow from her hand to vivify and fortify the vegetation. In exchange, she received energy from the plants that nourished her, in accordance with a Kosmic Voluntad (Will). With this emanation a relationship was established that nourished both. When Helena felt unbalanced, this action always helped her to calm her mind and harmonize her body so that the Ego could project itself more fully.

Once she had explained to Diego, *"**Plants have an emotional relationship with the persons that care for them, they have an exact and true perception of the environment and of the person with whom they establish said relationship. The plant perceives more than the human; this is because it has committed no crime. It is here to help man; it is an element of aid and comprehension. Any plant can cure us, if we ask its permission and help, because they have a transcendent energy"** (17 May 91).*

She walked upstairs to her room to prepare for the trip. She dressed simply, in a burgundy sweater, khaki pants and her favorite boots, ready to face the chilly day. She entered the kitchen and prepared a cup of coffee on the old cast-iron stove that helped warm the house. She was thinking about Diego and of the transcendent moment that he was living when she heard a knock at the door.

The trip to the caves in the Ardèche gorge was a short one; they were located a little more than an hour from Remoulins. Diego drove trying to focus so as not to mess up again. He was nervous, thinking of how to begin the chat with Helena. In that instant Helena asked him, "Alright, Diego, tell me. What's bothering you so much? Why are you so angry?"

Diego was taken aback; he didn't remember having told her about his anger, he had done everything possible to seem cheerful.

"Well, let's see, it's just that I'm really confused Helena. I don't know what I should do. I feel trapped with my dad." He halted again, trying to arrange what he was feeling so that it would sound alright to her. "What you said to me the other night about him and that I'm staying with him because I feel an obligation, took me by surprise. I'd never thought of it that way. I hope that you don't think I'm being selfish or that I'm a bad person, but it's that I want to do something different with my life and I don't want anything to interfere with it." Feeling a bit more relaxed, he continued explaining his feelings. "Look, the truth is that I have no idea of what I want to do with my life. I just know that I don't want to be confined in something that I don't like. Neither do I want what everyone thinks they should have:

kids, a nice house, cars, and so on. I want to do something, but I don't know what. You see, I can't even tell you what I want, just what I don't want. What do you think, Helena? You know that I like architecture, but I have to admit that deep down I feel that it would be just another dead end. I think I do a lot of things just because it's what's expected of me. You know, that idea that we have to do something useful. Oh! I don't know if I'm explaining myself," he exclaimed, frustrated.

Helena asked, "And how do you think I can help you, Diego?"

"To be honest, it's as if I were afraid to know what I want." He waited for her reaction, but she let him keep talking. "I know, it's weird, right? Why am I afraid of it?" He lost himself in his memories for a few seconds and then continued. "You know, the few moments of security that I remember were as a boy, when it seemed to me that the world was filled with mysteries and that I could go and discover them. I used to heal the animals that we had on the farm. Didn't I ever tell you about it?" Diego smiled, thinking about those faraway years.

"And how did you learn to do it?

"I just did it and sometimes it worked. It was as if there was a part inside of me that knew how and I let myself be guided. I simply put my hands on their heads and let my energy flow. I haven't done it   85 since then, but it's something that I'd like to be able to do again. But so much time has passed, that I don't know if I still have the ability in me. As a child you can do a lot of things, can't you? Why do you think that I could heal them, Helena?"

Helena analyzed what Diego told her while she watched the landscape through the car window. She only responded, "Tell me more. Why did it interest you?"

"It was an instinct that I followed; I only wanted to help them."

"And you've never tried to heal your father?"

Disconcerted, Diego answered, "Could I really? It never occurred to me, what an idiot! But do you really believe that it's possible to help my dad?"

Helena had found a thread to try to guide Diego towards understanding the reason that was behind what he was living. "Look, I'm going to tell you a bit about healing. But first, you should know that in order to cure an illness, one has to understand its origin and ask if it's permitted to treat it. Regarding your father, this is the first thing you'd have to comprehend. Well, the truth is that I don't like the word illness because it comes with the idea that it's something bad, that there's something that should be eliminated, but many times they are necessary experiences. So, we need to understand its cause in order to be able to treat it, if indeed that's what should be. In general, they are imbalances that result from an excess or deficiency of energy, or from the environment; although, there are also karmic reasons that provoke our maladies, and in the same way there are also the individual reasons necessary for the development of the Ego."

"Then, they aren't always caused by bacteria, viruses, or physical reasons?"

"It's more complicated than that Diego, but there are certain basic principles. To begin with, you have to understand that we are sonorous, the Ego as much as our matter, thoughts, and emotions. Anyway, sometimes we suffer variations in our sonority or our energy. This fluctuation can be the product of emotions, of the environment in which we live, of our own thoughts, or of those reach us from our surroundings. The reasons of our past lives which create habits and obsessions that remain registered in our matter also have an effect. Sometimes they manifest themselves in the present and they affect us. Everything converges in our present lifetime. *The pequeño (little one) has a quantity of energy that is its own sonority and it is in what we call matter. This quantity of energy forms a balance with the environment and with other reasons and when the pequeño suffers a decrease in sonority, what we call illness results, which can be due to the person does not eat what is necessary, is sad, etcetera [...] (19 Sep. 87).*

There are illnesses that come from mental reflection, because of obsession of personality, because of karma, illnesses that come when the Astral is filled with larvas and others when the larvas drain the physical matter" (18 Apr. 87).

Diego interrupted her, enthused by what she was explaining. "What do you mean by astral larvas? I haven't heard those words before."

Helena explained a part of what she knew: "All matter, thought or emotion generates a residue that remains in the environment. These wastes can form conglomerates or larvas. Sometimes the environment or the body becomes saturated with these residues and they affect us negatively. I don't know if you've realized that when you're in a place for a long time, you saturate it with your thoughts and emotions. Well generally, these wastes fall to the floor and they can get to the point of binding together and forming conglomerates or larvas. Many times when we arrive at a place we feel that it has a good or bad 'vibe,' and it's due to this reason. When a person is in harmony with their environment, he feels it; in the same way, a person with a lot of anger leaves these emotions fused in the site. It's important to maintain our matter and surroundings clean; the same as our thought. *Negative thoughts create monsters, and the positive ones, harmonious forms. Even more powerful than the images of thought are the forms that arise from feelings. Because of that, we must control our resentments, jealousies, etcetera. Our way of life must be clean, since the dirty particles create negative forms, as all matter is sentient (1 Jul. 87).*

"Now, we'll see that this same reason of astral larvas also affects us in a wider environment. The thoughts produced by the whole of humanity form something like a mass that envelops us. *Dense, impure thought* *forms a film. Thought is the atmosphere. The human body is formed in a certain percentage of thoughts, and thus also the cities. It is because of this that in the human conglomerates 'smog' is constituted, formed in part by toxic particles and in part by impure thoughts. The negative thoughts form a crust, for example: the frustration of the people at not being able to buy Christmas presents are thinking particles that float (12 Dec. 87).* These sounds, since thought is sound, saturate the atmosphere. The crudest fall, stick together and form astral larvas that affect matter, the mind and the Ego. By contrast, the most elevated thoughts rise to the high atmosphere, and are transformed into a brilliant lux (light)."

"Don't tell me that these sounds really affect our health?"

"Yes, that's why even our own thoughts can harm our health, or help us to improve. For example, if a person fills themselves with thoughts of anger and resentment, these are sounds that remain trapped in the matter. They form something like a 'cyst,' affecting the harmony that should exist in the body, and the person falls ill."

"Is that what happened to my father?"

"Well, there are many reasons for why we get sick. Do you remember that I told you that maladjustments occur within the ensemble of nine layers of our matter? Well, sometimes they fall out of adjustment and they aren't synchronized like they should be. They are thrown out of adjustment when we are disturbed by an experience, by our thoughts and emotions, or by factors in the environment. A delay or premature advance of one of the layers affects the rest. Even our memories and fears can provoke a maladjustment. *Our memories make us lag behind. Our anxieties make us get ahead of ourselves.*

We should always be happy in order to not get ill. When life is not understood, it is suffered, one is not happy and that is when we become ill (6 Jun. 87).

"We always live with the yearning to overcome the process of illness, to arrive at a physical perfection. In the past we had it; but we lost it when we didn't accept the matter we had and began to modify it. *In the beginning of the HTimes man lived longer and without so much suffering. The current matter is many sounds at different rhythms, like clocks that mark different hours, some fall behind and others run fast.*

The psychiatrist of the future will investigate the past. No one accepts themselves because of references to earlier times; for example, of a perfection that now they do not have" (12 Nov. 88).

"Wow." Diego was stunned to see that illnesses could have many causes that he had never imagined.

"Sometimes there are motives beneficial to the development of the Ego which provoke necessary changes and we perceive them as sickness. But in reality they are adjustments; processes that propel the Ego or the matter to an advancement. In this case, the illness is a modification, mutation, or preparation. *These states of the Ego, over which it has no control or will, are due to Kosmic reasons and acts in which it participates as part of a part (23 Jul. 99).* There are also karmic reasons, but I'll talk to you about them another day."

"I imagine so."

"Sometimes suffering is obligatory so that an Ego understands and advances, Diego. As I explained to you once, there is a Voluntad

(Will) that governs everything related to our development, and also for each particle in the creation, from the atoms to the great galaxies. Even the suns can get sick. Nonetheless, we shouldn't interfere in the reasons of another particle, whether it be a person, an animal, a virus, or a planet, without understanding if there is a Voluntad to do so and asking permission.

That's why there are so many reasons that we need to study first to know if we should or should not cure a person. And above all, know if the Konceptos give us permission to do it; if not, we end up disturbing the process that the Ego needs to live. All healing, whether through emanation, or with energy, which is what you did with your animals, is in reality an adjustment within what man calls 'gene.' Even though it's not expressed physically, it will provide an equilibrium between Ego and matter in future incarnations. This gives a harmony."

"And how can I know if I should heal my father?"

"As a first step, you should ask the Perespíritu that guides you if it is Voluntad to do it. The whole subject of healing is very interesting Diego, the same as the explanation for why you knew how to heal." She paused, looking at the stone cliffs and the river that ran below. "Look, we're arriving at a really interesting area. Why don't we stop here?"

Diego pulled off the road and parked the car. They got out, taking the backpacks with their cameras and food. Helena asked him, "Are you in the mood to explore for a while? There's a really interesting place a little to the east and there are lots of caves over there. While we walk, we can continue talking." With these words Helena started off. Diego wanted to know more about healing and wished that Helena would reveal to him why he had had the ability to cure when he was a boy. He put on his jacket and followed Helena, who was already descending the path to the river.

Chapter 6
Arriving at Destiny

To be true you must rid yourselves of the thirst for power and
generate humility. Be humble and you will be great.
ADVICE

Diego followed Helena along the path that went down to the
Ardèche river. They walked in silence between the huge rocks
and the trees that lined the edges of the ravine in which various caves
were located. The sinuous canyon had been formed by the activity
of the river millions of years ago. Since those remote times, the high
walls of limestone had been witness to the enormous changes that
the Earth had experienced and to the presence of the men who had
populated the area throughout the eras.

Helena observed the landscape and contemplated the process of
formation of the Earth, or Planicio Terráqueo. She was imagining
the condition and experiences of the Ego in stone so many millions
of years ago and how the Earth had changed since those first ages.
'What would the process of formation of the Earth have been like?'
she thought. She remembered what she had understood in the studies:

'When the Planicio Terráqueo was in formation, *at first it was an*
igneous mass. There were no colors other than black and white, it all
looked that way (Carpeta Dorada). The Entities responsible for the
creation of the earthly forms and with the supervision of the Planicio
Terráqueo gave shape to all that would be developed throughout the
Seven HTimes. *This New Evolution which was planned and sketched*
in a manner that would be coupled to the full understanding of
the Egos, since this was to be slow, without hurry, was so that each
achievement in the advancement would be a gift, and everything

was so (Carpeta Dorada). [...] It was in the moment in which the Virginal Waters arrived that the Planicio Terráqueo acquired the colorations that we now perceive. This was due to the fact that the rays of the sun were reflected in the vapors that the waters produced, generating in this way the colors of the rainbow, in which all of the colors that we perceive are contained (Carpeta Dorada). The vapors that were produced when the water touched the hot rock formed the atmosphere and this introduced the Egos into a kind of ensoñación.* (*A state between dreaming and waking.) Within this ensoñación they could begin to perceive and understand.

In the First HTime, that is, in the Astral, after the Earth was cooled, the Egos Deidades began to find in their first form of life, at once so ethereal and grandiose, the progress of their physical development. But the evolution in this Astral time was slow. With the slowness that was merited centuries and more centuries passed so that the Egos Deidades would have their first form of life, so that the Earth could shelter the Ego in this way of existence.

The Ego lived *in the rock that the water of the sea lapped against.* 91 *Afterwards in the water it was a kind of transparent cell with a point of apparent color, exhaling a trino (trill) of joy when it felt itself alive. [...]*

These conglomerates of cells scattered themselves and each one had a distinct sound, in such a way that different tonalities were heard which formed chords and arpeggios. These songs were already a harmony with the ruler of the Kosmo.

After, from this "cluster" of little Egos, there began to sprout in their growth diverse extensions, which gave rise to different miniscule little bodies, and from them arose an insect, an amphibian, a little egg which later would be a small bird, and from this transformation continued the metamorphosis of a small insect to another better endowed until arriving at those that we now know [...]. In the same way it occurred with the birds. From the metamorphosis of the

amphibian it went on until arriving at the mammal, of a very high evolution (Carpeta Dorada).

And so the Ego continued in its development, passing through many lessons in different animal matter. In the animal were manifested *certain perceptions and certain laws which ye have called instinct, the lessons penetrating in this state of the Ego without it being aware of this. Ye will understand likewise that the animal has great periods of calm, in which it regurgitates lessons that have been instilled in it. If it accepts them, it will pass therefore to the end of its experience as an animal, being prepared to be part of the human race, this being, so that thou understandest, like a prize, since it has understood the laws imposed by mandate and has assimilated the continuity of the manifest reasons (10 Jan. 01).*

In the earliest ages on the Earth, *the first men had matter very different than ours [...]. During the day he enjoyed three important periods: that of the morning, in which he fed on the manna that he received from the earth itself; that of the afternoon, in which he fell into a drowsiness, an ensoñación of delight caused by a manna that was like a blue cloud which enveloped him; and that of the night, in which he dreamed.*

That first man perceived the world with all of his body, with each particle of his being; it was not that he heard or saw only with specific organs, rather that he heard, felt, enjoyed with all of his body. In the present, some pequeños retain some of these excellences and can "read" a text with their eyes closed or "see" colors with the hands, excellences that should be recovered in the future. In the same way, the movements of those men were beautiful, graceful, like in a dance, since they were a response to the sound that came to them from the world; they perceived the sonority of silence. Everything produced a sound, and those pequeños perceived it, just like the great musicians, who called them the sounds or music of the spheres (27 Sep. 88).

The animal was the advisor and friend of man. That which startled man, that which man did not understand and did not perceive, the animal understood, and the communication between both was constant and projected itself when there was a profound friendship over great distances.

The first animals were fully intercommunicated with the Plant Kingdom and expressed in the stone their attitudes and instincts, the rocks being like a nucleus formed in the relationship of the ones with the others.

Man *was saturated with the Kosmic presence upon approaching the Plant kingdom; it is for that reason that man goes to the countryside to perceive something of what he had and was his nourishment'* (21 Mar. 83).

The clouds had scattered and the day was sunny and pleasant; Helena noted the change in the air that marked the transition between fall and winter. She stopped in front of some rocks and removed the camera from her backpack to take pictures. Diego took advantage of the pause to continue with the conversation that had so moved him.

"I was thinking about what you said to me in the car about my dad. I understand it, but it's difficult to accept. I don't want to see him suffer."

"You know what, Diego? You don't want your father to suffer because you have a strong connection with him, and it's not just from this lifetime. Families are people with whom we have many ties from the past, and the ties don't let us see things clearly," Helena clarified, while she focused the camera to shoot a picture.

"But we all have ties, Helena," he responded, nervous about what she was going to say.

Helena walked to the edge of the river, watching the leaves floating on the surface. Turning to look at Diego, she said, "Look, come and observe how the leaves follow the course of the water; some form groups among themselves while others find their trajectory alone. The solitary ones are able to flow more easily with the current; the groups of leaves invariably get caught in the rocks and branches along the shore. The same thing happens with us: there are people that look to join with others to pass through life while others prefer to do it in a solitary manner. Nonetheless, it's easier to find our destiny when there are no bonds or ties that impede our progress."

Diego stopped gazing at the leaves and looked at Helena.

"Then how should we relate to each other?" he asked.

"The apprenticeship of the Ego is solitary. When we committed the error in the Kosmo of the Trinak we separated ourselves from our relationship with the Konceptos and we began to feel an emptiness and the lack of a guide. Now, without this relationship, we're afraid of being alone because we don't know to where or how we should direct ourselves. The emptiness that we feel disturbs us and that's why we look for company, a relationship, a family, a job, to try to remedy this loneliness. That's why families, marriages, patriotism, or religion have been counter-productive reasons for the development of man." Seeing the question forming itself in Diego's mind, Helena added, "It's that the ties that we make with another person, with an idea, or with an ideology don't have the purpose of advancement on a Kosmic level. On the contrary, we form these ties based on emotions and for other unclear reasons. Sometimes it falls to us to have a relationship or a friendship, or live a circumstance with another person to have a lesson. But we want these relationships to last forever and we forget what's vital. Or we use them to reinforce our interests, whether we're conscious of them or not. If it falls to us to live something it's because we must find a lesson in it and we shouldn't try to make it permanent. Our life should be of constant changes and transformations. *An event, relationship, or circumstance is vital in a given moment for an apprenticeship and then ceases to be so that the Ego moves on to another motive of apprenticeship; it is for this a Real-Aparente (Apparent), Real because it has a transcendent reason to be for an evolution, and Apparent because it is only a projection conceived for that end (7 Jan. 89).*

Keep in mind, Diego, that when ties exist with a person, it's because we've made commitments in the past, for example, promised to love them or care for them, or even take revenge on them. They're reasons that remain inconclusive because they didn't arrive at an objective of transcendent improvement. Of course there are other reasons that are much more complex and that have to do with our condition here on the Earth. I'll give you a simple example. Let's say that in the past you agreed that you were going to take care of your father forever. Ok then, this commitment remained in you, and what you're living with him now is a repercussion of that programming. And I'll tell you something else: we're born within a family with the persons with whom we have many unresolved reasons, that is, karmic reasons."

Diego interrupted her, uneasy. "Then, the relationship with my father has roots in the past and now neither he nor I understand why? Wow, Helena, it's really complicated!" He picked up a small stone from the ground and tossed it into the river. It landed right on top of a leaf that was navigating alone in the current. "Oops!" he exclaimed, and blushed.

Helena smiled and shook her head. She noted how things moved Diego. She commented, "Talking about these things isn't easy, is it? Look, you decided to stay in Remoulins because of the obligation you feel towards your father. Maybe it would be better if you use this time to analyze the relationship you have with him. Or to see if maybe there's another reason for which you're here."

"Are you saying that my father shouldn't matter to me, Helena?"

"I'm just telling you that for each Ego to have the opportunity to search for its own evolution it must do so without commitments or alliances with other Egos. Its path on the Earth should be solitary, but it's going to learn by means of analyzing its interactions with others. It can *live affection towards others, but outside of it, without letting itself be affected and perturbed, so that it can be whole and project itself to superior levels (7 Jan. 89).*

That's why it's essential that we don't lose ourselves in reasons that aren't important, such as those duties dictated by our social and moral programming. How many people do you know that have to make a change in their life, but don't dare to do it because of their commitments, or because of their fear of being alone or of losing a material security? You need to remember that your father is just a man and in this incarnation he's with you in order to resolve something. Once you understand, you should continue with your own development. This way, you won't compromise your elevation because you feel a sense of duty."

"But that's being selfish, Helena. Wouldn't it be better to consider what my father needs as well?"

"You've had many fathers throughout your lives. I know that it's difficult now, but your father is only another Ego among the many others with whom you've been related, and you don't even remember them, right? Look, I don't mean to tell you to wash your hands and forget about him, rather that you separate your reasons from his and that way you'll be able to understand why you're together so that these reasons don't affect you. The important thing is that you don't involve yourself in reasons that don't belong to you. We shouldn't interfere in the evolution of another person; the effort of each should be comprehension.

There's something else you need to know. Everyone is born with a 'package' of all that corresponds to them to have in a lifetime. If a person is born with assets, health or beauty, it's so that they use them for their development, not to have a comfort in which they lose themselves. If a person is born with problems or disadvantages, it's also so that they surmount them. For example, if you're born poor, it's not a matter of simply finding a way to earn a lot of money. Maybe it's an experience that is necessary in order to understand that the material isn't of vital importance and that the Ego should continue developing even without certain advantages. Now we arrive at an important point. If you continue with your father in this life without understanding why you're doing it, this reason remains inconclusive again and you'll have to repeat it until you do understand. If you only remain with annoyance or anger because you can't do what you want, you don't advance either. When a person asks for the opportunity to break with a tie or with a karmic reason this is suspended as long as he fulfills what is vital." She remained silent for a few moments and then continued talking. "Look, Diego, let me explain something to you. When we form ties with someone, we distance ourselves from our own reasons because we confuse ours with theirs, with things that don't correspond to us. *If an Ego rotates around another Ego, it loses itself, it becomes negative [...]. Thus man should rotate towards the reasons that gave him life, this is paramount, but he cannot do it because he does not know* the Kosmic reasons" *(27 Sep. 88)*. Helena stood up and began to walk.

Diego grabbed the backpack with his camera and said, "You know what? I'm going to stay here to take some pictures. I'll catch up to you in a bit, ok?" Diego recognized that she was right, that he was stopping himself because of his father. He wanted to be alone for a moment to think about the things that she had told him. He was annoyed, uncomfortable and didn't know how to square Helena's words with what he had learned over the course of his life. He felt as if there was a huge clash inside of him.

Helena followed the course of the canyon, taking pictures of the limestone formations. A while later she heard Diego's footsteps behind her approaching rapidly. She turned to see him; he had mud on his pants and on his face.

"What happened, Diego?" she asked, concerned.

"It's ok, go on, Helena; it was nothing, I just slipped trying to take a picture on the cliff," responded Diego, shrugging his shoulders so as not to give it too much importance. Helena smiled and shook her head, seeing Diego's distress reflected in his little accidents. Starting up where they had left the conversation earlier, he asked her, "I understand what you told me. But tell me, how can I know what I need to do with my life? If there's a path, how do I find it?"

Helena said, "Do you remember what I told you about the Perespíritu? *The motive of the Perespíritu is to put before the Ego the reasons that are vital for it and preserve the reasons of the matter in its roaming within the structure of the sphere (1 Oct. 99).* Throughout the ages we have perceived it in different ways, for example, what some people call the voice of their conscience, or like a guardian angel, although it's not what people refer to as angels, it's one of the layers of our matter. Why don't you ask it about what you should do?" She turned and invited him to keep walking with her.

"Yes, I remember that you talked to me about it." Diego remained silent. He followed her for a while, thinking and looking at the forms of the rocks carved by the water and the air. Later on he commented, animated, "You see, when I'm alone and calm, I'm able to perceive it. Sometimes I feel impulses that indicate to me that I should or shouldn't do something. You know, you always feel it when you screw up. But it's hard for me to understand it well when I'm really emotional. Does this have something to do with what you do? You've told me that you study and analyze; do you talk to your Perespíritu?"

"No Diego, I speak with the Konceptos (Supreme Entities), or what created us. And the person who introduced me to this was Marion and she in her turn by the Great Maestro. For you, Diego the interrelationship with the Perespíritu would be like a first step. It's a principle that everyone can and should develop. But, why don't we continue talking about this later? I'm very hungry, what do you think if we stop to eat?"

They sat on some large rocks to eat their baguettes. Diego took advantage of the stillness of the moment to ask Helena something.

"There's something that really interests me in what we were talking about. How did you come to know about everything that you tell me?"

"You don't know this, but many years ago I was studying medicine in Paris."

"Really? I can't imagine it," exclaimed Diego, astonished.

Smiling, Helena responded, "Neither can I. Well, we all have our weaknesses. I went through many experiences, trying to find something or someone that could answer the three questions that I had: who am I, where do I come from, and where am I going to. The years passed and I knew that in order to find my path I would need to be alone; I knew that as long as I lived in my parents' house I wasn't going to be able to advance. At that time I wanted to help people, I saw the suffering in the world and thought that I could do something about it. And so I decided to study medicine. I really enjoyed it and I felt that finally I had found something in which I could be useful.

Doctors try to do good deeds and try to help, but many times they don't cure us. Why? It was a question that fascinated me. Why was it that one person was cured and another not, or why did one patient fall ill from a hereditary disease but his sister, with the same genes, didn't? It was something that I hadn't been able to understand. I was almost at the point of completing my degree. I had studied a lot, but I realized that not everything was found in books. Many times I used my intuition to diagnose the patients. I didn't understand in that moment how it was that I had the capacity to perceive what was producing their symptoms or how to alleviate them. And I was piqued by the question of what was the origin of illness in principle.

Well, it was in that moment that I met my Maestra; she explained much to me about illness and about lots of other interesting things that no one else knew. I began to study with her and, among many other things I arrived at understanding the human body from another angle. For example, scientists only understand one part of it and that's why they only see one aspect of the process of illness and of healing. When I studied about the Ego and our reasons, I understood much more about the origin of the maladies that present themselves in us. The matter should help the Ego in its transformation, but normally it's the other way around; the Ego is subject to the demands of the matter and from there stem all of the imbalances that provoke our physical and mental disorders."

"Couldn't you have continued with medicine at the same time, supplementing it with the other things that you've learned?" Diego asked her.

"No, Diego, I had to leave that reason and all that my life had been before. When a person is initiated in this road, he dies to his previous life to be born into another more complete reason. Besides, medicine is a reason that is based on Wisdom and what I study and live is Konocimiento. They're two distinct reasons. But my life is another story. What you should understand now is that many times medicine isn't able to cure us because as a first step we should treat the Ego and not our matter."

They finished eating and decided to take one last hike. They climbed up to the rim of the canyon to have a panoramic view of the area. They walked along the path surrounded by the sounds of the place: the wind, the songs of the birds, the water that ran in the river. Diego realized the he was starting to feel lighter and had more energy. His mind was clearer.

They arrived at a high point and both of them felt as if they were in another time. They took some more photos and Helena commented, "Look, it's clouding up again. If we want to get back to the car before it starts to rain we'll have to hurry." They walked back to the car. They set off and drove a while in silence until Diego dared to ask something else of Helena.

"I wish I could do what you do, Helena. But I don't have a preparation like yours, my life and my experiences have been really simple. Even if I'm not prepared, do you think that I could learn about the things you tell me?"

"What I learned in school didn't help me in these studies. True intelligence isn't an education within human reasons, it's something else. In the studies we see that *the intellect is understanding the vital and this is the ELEVATION that is Evolution.* The duty of the Ego *is to understand itself and accept itself, and to put a definition between what it is and what Matter is. What should be its own comprehension? To accept itself as a particle of a "WHOLE" and to have the submission and the impulse to understand its relationship with what has manifested it (12 Jan. 09).*

"I want to clarify something for you Diego: you don't have a life full of material reasons in part because you chose it to be so. You've had a spiritual development before. To arrive at the point where you are now, open and with the desire to learn, speaks of a sequence in your lives. In them you've been searching, because you've realized that there is something more beyond what you see with your eyes. That's why you know a little about healing and you have a rejection of getting tangled up in human reasons. If you analyze it, you'll see that you have a certain freedom, except for the old man that you chose to take care of." Diego gave a slight start. "But that's how it is, right?" Helena said at seeing Diego's reaction to her words. Fortunately, Diego laughed, letting out a guffaw and freeing himself from the taboo inside of him. She continued, putting her hand on Diego's shoulder. "What's happening is that you're neither here nor there. And you're stuck in a hard place now. You have to make a decision. Are you going to continue as you have been or do you want to advance?

Humanity has a very specific opportunity right now. The Konocimiento Kósmico has been delivered so that we can understand who we are and the relationship we have with the Universe that we inhabit. To take the next step in our development we have to free ourselves from all of the ties that we have formed. Only in that way can we find our destiny."

"And if I wanted to study, what would I have to do?"

"I don't know, Diego. What do you offer in order to receive the opportunity? The Konocimiento isn't a course that you take in school; it's a path for all of your life and for all of the following ones."

"You've told me that you left your career because you couldn't continue in those two reasons because they would end up opposing each other. If I think about it that way, I would have to leave my attachments. And so I would need to leave…" Diego couldn't finish speaking. The words stuck in his throat. He continued driving in silence, focused on the decision he had to make.

Chapter 7
Mexico

Why is the water clean and clear if we insist on dirtying it?
Because it is Eternal.
The same water that arrived is that which will leave.
KONOCIMIENTO

The streets full of cars that circulated rapidly and the sound of horns showed off this enormous city. With 28 years of age and eager to see all of the changes that this trip would bring, Helena arrived at her hotel in the center of Mexico City in the summer of 1996 at 6:30 in the afternoon, the busiest time of the day. It was the hour when all of the shopkeepers in the downtown area closed their businesses and the workers returned to their homes. After nearly two hours on the road, the taxi had finally left her at the door of the hotel. In the time in which the driver took to drop off her suitcase, more horns honked than Helena had heard in her entire life.

She arranged her suitcase on the bench in her room and took out the few clothes and belongings that she had brought. Five years had already passed since she had suddenly left her medical studies. She had been working as a waitress and doing small photography jobs, earning enough to be able to live very simply in Paris. She didn't have many possessions, except for her books, but she had known there would be no sense in taking them with her, since she had no idea of what her life would be like from this moment on. Upon being near her Maestra Marion, her life whirled dizzily, full of happenings and unexpected situations. She sat on the bed and thought of the speed with which this relocation had developed. In the last two weeks she had to leave her job, her apartment, and get rid of the few possessions that she had. She made the trip as quickly as possible. She had learned from her experiences that when she had to do something transcendent, she should act with haste because once an idea was projected and began to develop itself, divergent thoughts arose that impeded her progress. Helena was surprised to understand that these discordant ideas could come from the environment, from the people closest to us, or most of the time, from oneself.

Now she was seated in the hotel, in a place that was strange to her. Tired, but with the elation of being able to see her Maestra Marion again, she called to let her know that she had arrived.

"Well, you finally got here, Helenita. What took you so long? I thought that you were never going be able to break away from your comforts. The croissants are really good over there, right? Well, I imagine that you must be exhausted." Helena recognized her Maestra's ironic tone. She always found her weak spot. "Ask in your hotel for the restaurant El Moro, it's close to where you are. We'll see each other there at eight." Helena hung up the phone and looked at the hour; she took advantage of the little time she had to rest a bit. Her Maestra always made her nervous and she wanted to be in harmony as much as she could be for the meeting.

When she arrived at El Moro, Helena couldn't help but smile upon seeing what they served in that restaurant: *churros* (crullers) and hot chocolate, something that of course her Maestra couldn't resist. She saw her at one of the tables close to the entrance and approached her, feeling a great sense of contentment at finding her again.

Marion greeted her, speaking in Spanish. Helena sat opposite her, observing the tiles on the walls and the ambience so different than that to which she was accustomed. She ordered a hot chocolate, seeing with fright the plate full of *churros* on the table. She hoped that she wouldn't have to eat all of them. Marion began to ask her about her trip. At first, Helena was a little doubtful of not being able to follow the conversation, but she managed to concentrate and flow. Following Marion's advice, she had studied the language in Paris; this would be the moment to put it into practice.

Marion sipped a little of her chocolate and said, "You should accustom yourself to speaking Spanish; perceiving its sound will make it easier to gain an appreciation of the people and their way of thinking. Besides, the Konocimiento of this HTime is written in Spanish. In the future only one language will be spoken and it will be based on Spanish and Macedonian." Marion smiled at seeing the expression of surprise on Helena's face.

A rhythmic and repetitive sound reached Helena's ears. Outside of the restaurant there was a group of *conchero* dancers dressed in prehispanic garb. The form of the dance made Helena think that it was a ritual. The rhythm of the rattles tied to their legs and that of the drums provoked a heaviness in her mind. She wondered, 'why do I feel this way?'

Knowing what Helena was thinking, Marion commented, "They're rituals. Their music and their habits have been repeated for a very long time. If you examine the customs of the native peoples, in their thoughts and even in their dialects, you'll see that they're obsessive and repetitive. In the past, these races had a moment of splendor, but when they didn't carry out what they had to do, they suffered a decline and their konsciousness was limited. They're like scratched disks that repeat and repeat their programming without getting anywhere." Marion explained in more detail. ***"Humanity has had many prototypical races in the ages gone by, and each race harbored a set of qualities and wonders that didn't reach their total goal. From these same races there was a remnant that remained, but without the perception and amazement that upon having been chosen, they had had in the past. This difference that is subject to their color and their features, as well as to their biological and psychic functions, and their attitude towards life, will remain imprinted always in the subsequent races, which will continue their transformation until they complete their ritual of instruction and advancement in their evolution. All of them in a propitious and indicated moment could have arrived at the point that had been marked for them, but it was not so. Each one was lost in dark and personal interests in each age (12 Jul. 96).***

The races that lived principally in Mesoamerica and in other parts of the world emerged from other more ancient civilizations. In the past there were other civilizations like Atlantis and Lemuria, but neither were they the greatest and most developed races; they were already not at their zenith. Part of their wisdom was an

interpretation of the Konocimiento they had from the earlier races that were in plenitude. Of these races, those men who were saved from the cataclysms that occurred departed to different places in the world and remained principally in America. They became Toltecas, which does not refer to a race, rather to those that wandered everywhere, teaching that which has to do with what is known as the Astral (18 Nov. 08).

The Konocimiento that today we understand as Astral was taught by the Maestro of humanity to the first races, which in that moment were in plenitude. It was given to them with the intention that they would understand their matter and the Planicio Terráqueo, so that they could have a communication and harmony with the Entities charged with the functioning of the earthly sphere.

These races made great strides in the lessons of the Astral, but with their interest focused on recovering something from the past: to gather together their former hierarchical leaders."

Helena, surprised, turned to look at the *conchero* dancers that moved in an almost robotic way, tracing figures on the ground with their feet and following the repetitive rhythm of the drums, which made her sleepy and seemed to put them into a trance. She understood that this rhythm served to focus their attention.

"Do you remember, what do I mean when I talk about hierarchical leaders?" Her Maestra kept looking at her and paused so that Helena could remember what she had studied. Without waiting for Helena to answer, Marion continued. "I've told you previously that the Egos were formed in Huestes (Hosts) in the Kosmo of the Trinak. Humanity is conformed of the same Huestes, each one *with a certain conditioning and capacity to develop the variety of missions; it is therefore that men are laborers, workers, thinkers, scientists, etcetera. This attitude of the Huestes is for a necessary end (14 Mar. 92).* Each Hueste was guided by Egos that had a greater preparation and a more elevated konsciousness. These 'senior' Egos should have helped the 'littlest' to develop, but it wasn't so. At the moment of the error in the Kosmo of the Trinak, the most developed had a greater liability, since they had more konsciousness of what they were doing. Their graces and great capacities weren't utilized for the assigned purpose, but rather to promote and participate in an error. We already know that after committing these mistakes, the Egos went through a Kosmic trial and were projected to the Planicio Terráqueo, where the new arrangement that they would have was decided. To achieve this, the triangular structure and the hierarchical order that had existed in the Kosmo of the Trinak were modified."

Her Maestra Marion adjusted herself in her chair, turning herself to face sideways and rubbed her brow with the edge of her palm. Helena observed this characteristic gesture that she made when something was worrying her. The waiter brought the hot chocolate to Helena while Marion took a pen out of her bag, and on a napkin, drew a triangle.

"Look, we can say that in the upper part were the Egos that should have guided the Huestes in their efforts." She emphasized the point of the triangle and continued. "Also, we should understand that there were entire Huestes that had a greater development than others. We can visualize it as if there were layers or distinct strata of konsciousness within this structure, with the more developed Egos in the upper part and the less developed ones below.

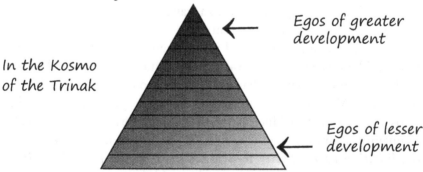

Upon arriving at the Planicio Terráqueo a new order was sketched, since the previous one hadn't functioned. The Egos that had been at the top of the triangle before were directed to the depths of the Earth." She turned the napkin over and drew on the other side an inverted triangle and signaled:

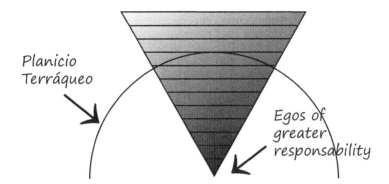

Planicio Terráqueo

Egos of greater responsability

"And so these Egos would remain in latency in the interior of the Planicio Terráqueo for a much longer time, since their interests and conduct required a long period of introspection and isolation, and in this way they would have the chance to fully comprehend their errors. According to how the development of the Egos played out on the Earth, possibly in the Sixth HTime, these would have to leave their detention to integrate themselves in the new form of evolution and instruction. Thus, the Egos that remained in the upper layers of the Planicio Terráqueo," she signaled the base of the second triangle, "would have the opportunity to adjust and find their place for themselves, without the interference of their former leaders, since the less developed Egos were accustomed to following them. It's similar to how jails function now: the most dangerous men that have committed the worst crimes are separated in cells or special prisons, completing longer sentences than those who have lesser faults and can more easily rectify their mistakes.

This equilibrium, which had been planned by the Supreme Entities so that the Ego could find its place and could understand, was broken when a few men, desirous of having more power over the Earth, wanted to unite themselves once again with their superiors in order to restore the arrangement of the hierarchies that had existed in the Kosmo of the Trinak." Marion lit a cigarette without realizing that she had another already lit in the ashtray. Her faraway gaze seemed lost in the ages. "But not even they comprehended the seriousness of the consequences of their actions. When they managed to remove some of these Egos prematurely from their isolation in stone, these came out crazed and furious, seeing as they had not passed through the necessary transformations that would have allowed them to understand their failings and their conduct. Filled with rage in the face of the new arrangement, since they had been stripped of their graces and powers, they sought at all costs to dominate the Egos of lesser blame that had already found a certain harmony on the Earth. These Egos that came out ahead of time *are what people call demons and are Egos that are in a great misadjustment since their error was greater. When they are liberated by means of a false, erroneous process, they are the great criminals or defenders of erratic or confused causes (11 Jun. 88).* These Egos imposed themselves upon men as their gods and made them worship them. Haven't you ever asked yourself why some of the 'gods' of the past are grotesque, demonical figures? The desire of these hierarchical leaders was, and still is, that the Ego doesn't perceive either its truth or the Supreme Entities. The hierarchical leaders want to maintain the konsciousness of man occupied in a world of material fantasy driven forward by themselves, in which they saturate the mind with greed of having it all: money, health, beauty, love, power, intelligence. In this way they dominate the rest while they remain at the top of the triangle. They direct the thoughts, emotions and actions of the submissive Egos, and they feed on these reasons, becoming more powerful and reinforcing the illusion of those that they maintain as their subjects."

In that instant a waiter who was carrying a load of plates bumped into their table and made the chocolate spill out of their cups. Helena was annoyed by the interruption; she wasn't accustomed to the disorder that she found in this place.

"Concentrate on what I'm saying and don't fuss. Can't you control your matter? You're like a little girl," Marion exclaimed and continued explaining. "With these actions the harmony on the Earth was broken and there were many dire consequences for the Egos. One of them is that the Ego, upon detaching itself from its path, stopped receiving the sustenance of the Konceptos. Then, the needs of matter dominated the Ego, which was left diminished and disturbed. Matter is necessary so that the Ego understands on the Earth, but the acts which were inappropriate, such as having eaten the animal, wreaked havoc on it. Matter now is a glutton, it always wants more; it always seeks comfort and security. Matter in this state is like a drug that anesthetizes the

Ego and doesn't permit it to develop its spirituality. And so the Ego can't comprehend its reasons, since, we'll say, it's asleep. Those that impose these reasons know that a man who has his mind occupied in what he thinks is necessary for his life, can't understand or sense his truths."

"Then those who have imposed these reasons on humanity are the 'leaders' of the past?"

"Yes, they're the evil ones that run the world. They're deranged minds that only want to subjugate their fellow beings; they're the wealthy who accumulate millions and millions that they can't spend in a lifetime, they're those who seek to leave the Earth in spaceships, those that subject the majority by means of fear or an ideology, those that provoke the wars in order to destroy the Planicio Terráqueo. There is no repentance in them; their arrogance is too great to accept it.

Look, it's easy to see an example in today's world through the idea of royalty. This idea is a reflection of what happened in the past when they wanted to restructure the hierarchies again. That's why 'your highness' is still said; it's a memory of the elevation that the leaders of the Huestes had in the distant past. But it's not the same now; these people don't have elevation, but rather power. Royalty is an example of a structure formed to maintain power within a handful of people who inherit for generations and ages. The royal families of Europe, for example, although they are in different countries, are almost all related and have inherited between them the money, properties and power that have permitted them to maintain this position throughout centuries. We can see this arrangement in many levels, not just in royalty; we just have to look at the structure of the economic, political or religious world. And mankind accepts this imposition because of the indolence of centuries and also because this structure forms part of the programming of the Ego to follow the instructions of a hierarchical leader. Nevertheless, in the future this is going to change and the structure of the countries and of the races will be based on the spiritual elevation of the Egos. The most elevated will have the responsibility of helping the rest to comprehend and evolve, as it should have been since the beginning. The races will be divided according to their evolution so that the less developed Egos, and those who have not wanted to accept the lessons of the Planicio Terráqueo, don't hinder those that do want to comprehend.

Tell me, Helena, if an Ego has understood what Life is about, why does it want to rule over another or over a country? Why does it want to have a hundred cars, many women, twenty houses, or a weapon that can destroy everything?"

Her Maestra remained silent. Helena perceived the disillusionment in her. She looked around; she felt a great oppression. Suddenly it seemed that everyone in the place had sensed Marion's words and had been bothered by them, since in that instant the restaurant became saturated with jarring noises and people gazed at them with mistrust. There was much movement in the crowd, children crying, *concheros* dancing, beggars that wouldn't stop approaching to interrupt them. She realized that the presence of Marion provoked anger.

The experiences she was having at this moment were new to Helena. Upon seeing the aggressiveness of the place, she asked, "Don't you want us to go somewhere that's calmer?"

Marion turned and said firmly, "I'm within myself, it doesn't matter to me what the surroundings are. Don't start with your prudishness now, or do you already want to go back to Paris?"

Helena adjusted herself in her seat, feeling the lesson in her words. She saw the drawing of the triangle on the napkin and understood the balance that the Konceptos had marked so that the development of man on the Earth was complete. She saw how man had distorted it, causing a great damage to himself.

After various cups of chocolate and a goodly quantity of *churros* that her Maestra had obliged her to eat in order to test her submission, they went out onto the street to stroll. The dancers had left. The long walk helped Helena pull herself together; now she felt perceptive and for some reason alert. They reached the main square; it was an enormous esplanade where they bumped into same group of dancers once again. They danced, forming two circles that touched at one end, forming the continuity of an infinity symbol. In this instant, Helena noticed something odd; she perceived something like a hole in the ground, from which emerged filaments that captured the energy that the dancers delivered. She thought of what her Maestra had just told her about the rites of the indigenous peoples and of their purposes. Helena understood that these dancers were here because of the nearness to the prehispanic constructions that were buried below.

She approached one of the *conchero* dancers who in that moment wasn't participating in the dance and asked him as a curious tourist, "Could you tell me what the meaning of the dance is?"

A little annoyed, but demonstrating great pride, the man answered, "It's a dance to honor our ancestors; we seek to awaken them so that they take their memories and their power." Helena pulled back at hearing this. "They are customs that have been transmitted father to son and that date from a very distant time. Are you both Mexican?" he asked Helena, fixing his attention on Marion.

"No, well, yes…it's that I have family, but…"

When Marion saw the disturbance and Helena's embarrassing moment, she jumped in to answer the *conchero* in a teasing tone, "No, I don't lift my leg to pee."

"Don't offend my country, señora." He pulled himself erect at hearing her.

Upon hearing Marion's remark, Helena struggled to contain her urge to laugh; tears started running down her face. But on the other hand she was stunned by all that she was witnessing in this place. She contained her emotions and responded to the indigenous man, "If you knew that she only says it because of the anguish and desperation that she feels at seeing this moment. In spite of the enormous care and effort she has put into her teachings, mankind remains the same in its foolishness of persisting in repeating an error by continuing with these rituals. Then you'd understand that she says it with a voice of great love and desperation at seeing that mankind has not understood."

In spite of the sparks that leaped from his eyes, the man remained silent, disconcerted, without fully understanding what she was saying. Helena turned and went running to catch Marion, who as always, was walking rapidly and had already left her far behind.

"What is that hole that I saw in the plaza?" Helena asked when she caught up to her.

"It's an outlet of energy from the Earth. Look, the Planicio Terráqueo has many places where these energies enter and exit, and they have been called '*bocas*' (vents). There are 'ducts' through which energy runs, and they cross the Earth everywhere; we can say that they are similar to the veins and arteries of our body: some recover the wastes and others deliver a force. It's by means of these ducts that the Earth receives the sonorous energy from the sun and consequently man is also nourished by it. This has been his principal food since the beginning of the HTimes, before he ate meat, since his matter was not programmed to feed on this.

Not all of the ducts are active at the same time; some are static and are activated over the course of time and those that are active 'go to sleep;' all of this is according to the how the unfolding of the HTimes will be. The energy that arrives at the *boca* exits in order to activate and foster a development of the Earth and in man: it helps to activate, among other things, his konsciousness."

"But they're using this energy wrongly!" exclaimed Helena.

"The energy is neither positive nor negative. It's delivered for an advancement, but it depends on man as to how it is used. What happens is that the Egos that are inside of the Earth know when a duct will be activated and they want to take advantage of the energy in order to escape. It's because of this that these characters with their rhythms, sounds, and movements provoke a reverberation that goes to the interior of the Planicio Terráqueo and impacts in other places that are connected by the same ducts. People flock to the sites called 'magical' to fill themselves with energy, but what they don't know is that when they do so, they send their energy to the Egos that are still in the depths. Upon delivering this energy they are helping them escape. The Egos in the depths are the devils of which humanity has spoken so often throughout the HTimes and are the famous gods that they adore. As the man clearly told you, they want to awaken them; that's what they seek.

Some of the inhabitants that participate in these rites are conscious of it and they are the guardians of the place, and others are their followers, or as they say here, their *achichincles*, most of them indigenous people. That's why recently it's become fashionable to go to these ancient sites to conduct ceremonies. Not just here and not just the natives. These practices occur all over the world and are carried out by people of different beliefs, but they all have the same goal. Here in Mexico, they come constantly to make their 'visits' to the temples and pyramids."

"*Achichincle*? What does that mean?"

"Serf, servile, slave."

Helena accompanied her Maestra to her apartment, thinking about all that had happened in such a short time. She remembered the studies that Marion had given to her on her last trip to Paris; they were like loose points that she was only now beginning to put together.

"We're getting close to my house," she said, taking her by the arm to turn to the right. Helena read the name of the subway station they had just passed: "Pino Suarez." They turned on Netzahualcóyotl Street, where there was a small plaza. It was now very late and there was almost no movement. Her Maestra stopped in front of the door of an old building, but even more than that, it was dirty and mistreated. It was the only one on this block that had remained standing after the earthquake of 1985.

"Your hotel is a few blocks from here." She explained how to get there and added, "You hadn't imagined so many surprises, right? Just wait to see what's coming."

Helena, nervous, wanted to laugh but saw the serious look on Marion's face. They said goodbye and she went to her hotel.

Of all that she had lived that night, what most distressed her was seeing the building where her Maestra lived. How was it possible that she lived in that place? She wondered why she would have chosen it. There must be an important reason for it. She knew that for Marion, a clean and well-cared for place was vital. She was terribly annoyed when she saw how people destroyed things, or that they didn't take care of them. With these fine points man demonstrated his urge for destruction. In these small details her Maestra saw the lack of respect of man towards the Konceptos, since matter, the Earth, the sea, or even Wisdom, are a loan from the Konceptos to the Egos.

Upon arriving at her hotel Helena was exhausted; there had been too many experiences in so few hours. She took out her notebook and jotted down all that her Maestra had told her and what she had seen in the plaza. She had much to study. She closed her eyes for an instant and the memory of how she had arrived at this moment came rushing back to her.

Helena had met her Maestra five years earlier in Paris. Since even before that, Marion had travelled frequently to Mexico, since this place was indicated for the delivery of the Konocimiento of the Third HTime. There she had formed a study group, of which at that point, Helena had known nothing.

When her Maestra had explained, on her last trip to Paris, about these reasons, Helena had asked her, "But why in Mexico? How is it that a place decided upon?"

"It has nothing to do with a country or a nationality. Look, it's really a Kosmic action. First we have to understand what happened in the past in order to know why this place receives this opportunity again in the present. We'll talk of the races that in the past lived in the Americas and principally in Mexico. They had great advances; they achieved a development within the Astral reasons that are now known as magic. As you know, magic comes from the teachings that mankind received in the First HTime to be able to understand and utilize the energy of the Earth and the planets for his evolution. With these he achieved great wonders, which turned dark and misguided upon using them only to have power and modify what had been established. That's why as races, they made a great karma for themselves. Like all instructions that are not completed, they need to be repeated until they fulfill their purpose. In the present an opportunity is granted once again in order to clarify that past. As some might say, Mexico is a representative in the eyes humanity, because what happens here will impact in all of the Planicio Terráqueo." Her Maestra paused and then continued. "As I've explained to you before, all of the events of the Planicio Terráqueo are related with each other. In this time some of the energies that nourish the Earth are active in the Americas. It's the mechanism of the changes in the Earth itself which are related to the HTimes of the Planicio Terráqueo and of the Kosmo."

"And how are these energies activated?"

"By means of the sonorous ducts of the Planicio Terráqueo. That's how the energy arrives in one specific moment and to a specific place that is signaled by the Supreme Entities in order to fulfill a marked action. This is the case of Mexico and the Americas in the present moment. In another time it will be in other places. In this new stage in Mexico there are two reasons that exist: on the one hand, the greatness of the Konocimiento that is being delivered, and on the other the insistence of gathering together a force between men in order to attract and release the rebellious Egos before their time."

Marion had commented to Helena about a study group that she had assembled in Mexico. She explained that they were making plans to search for a place so that all of them could gather around the studies. The idea that they had was to retire to a site where they wouldn't have the commitments and difficulties that apparently were what distracted them from studying. Together they could share what they learned and help each other to access the memories of what had led them astray in the past so that they wouldn't repeat it again in the present. They wanted to form a base that would serve as a support for their future development.

Her Maestra had told her that mankind along with the Earth were entering into the reasons of the Third HTime, which had begun a few years earlier, and that compulsory changes would have to come to mankind and to the Planicio Terráqueo to secure this transformation. She explained that these changes were not isolated events, but rather they were part of a constant transformation in all of the reasons of the Fusion (Universe), because all was connected. In this Third HTime the changes would be noted mainly in the Sun and Jupiter, which would enter into great activity. Helena wondered, 'what do these changes mean?' Marion gave her a study to help her understand more exactly.

'In each determined period within the reason HTiempo (HTime), [...] absolutely all is modified in the Campo Creativo (Creative Field) or Fusion. In thee pequeños terruales (little ones of the Earth), thy structure is modified in its totality, and thus, in what thou callest Konsciousness, and of course, in the mental reasons. In the Earth the internal currents of thy mother, which is the Planicio Terráqueo,

are modified. Many of the modifications are not visible, but they exist." She explained, "We have used the word evolution to be able to understand the changes that occur. *When this happens, new governing Entities enter into the Planicio Terráqueo and humanity should accept the new transformations, parting with various that were in the past*" (26 Jul. 00).

Helena had heard that people talked of a "new era" of "changes in consciousness," but without fully understanding why they were happening or the consequences that would result from them. Without this comprehension, man would simply be dragged along by the changes, like a wave that carries away all that it finds in its path. It was the moment to make an effort to understand.

Two weeks after her Maestra had returned to Mexico, Helena found herself washing dishes in the cafe where she was working. She began thinking of what Marion had explained to her a little while earlier about the studies that had been in other places in the world, but now were centered in Mexico. Helena realized that she had been annoyed by what her Maestra had told her about the happenings in Mexico, since for her it meant a change in her life that she didn't want to make. She didn't feel any attraction for going to another place. In Paris she was comfortable and becoming increasingly more successful with her photography. Then she remembered her first day of study with her Maestra and the commitment she had made. She realized that she mustn't stop herself now because of a comfort and a preference. She understood why her Maestra had constantly called her "spoiled little girl" and had made fun of her because of it. Little by little she had understood the importance that this moment had for her and for humanity. She knew with certainty that if she wanted to continue studying, she needed to leave for Mexico. She knew that she had been avoiding making the decision to leave her life in France because she didn't want to accept a responsibility. She put herself to the task of making all of the necessary arrangements to leave France as soon as possible.

On the afternoon of her third day in Mexico, Helena presented herself at her Maestra's house to study; that day she would meet the group. She barely recognized the place that she had seen that first night. By day the street was saturated with people, food stands and a public bus terminal. On the other side of the road there was a small plaza that was full of street vendors. In the entrance of the building a woman was selling greasy fried food to the passersby and to the bus drivers. With a high-pitched voice, she didn't stop shouting for one minute,

"Come on in, come in and eat your *gorditas*!"

With difficulty she managed to slip between the people to enter the building where Marion lived. In the hallways the children ran playing with balls. The music and the loud voices that issued forth from the various apartments blurred together. The floor of the entrance felt sticky. 'It must be the oil which that woman who sells the fried food throws out,' Helena thought. She felt a very deep uneasiness that was exacerbated by the mental filth that surrounded the place. 'I can't believe that my Maestra lives in this place.' She walked up a spiral staircase to the second floor and knocked on the door. Her Maestra opened it and greeted her.

"Enter, Your Worship," she said imitating the manner of speaking of the masses.

Helena didn't understand what that meant, but there was a hint of irony in her tone of voice. In the instant that she entered the apartment she felt as if she had passed into another dimension; the contrast was that great. The apartment was impeccably clean and full of light, a golden light that filled every corner and made her feel small. She had only perceived this glow when she was near her Maestra. A red rug defined the area of the living room, in which there were three sofas and a table in the center. On the white walls was hung a large collection of masks that were painted in a very peculiar manner. Observing the brushstrokes and colors, she thought, 'surely she painted them.'

"They're the neighbors," Marion indicated.

"I think they're identical," Helena answered, laughing. They were masks of grotesque faces that resembled demons.

Helena smiled at remembering the small fragments of that old dream that she had had a couple of days before first meeting her

Maestra in Paris, five years earlier. In that moment she understood that the building she had dreamed of was her Maestra's house in Mexico. She was surprised and remembered a little more of the dream: the dirty and mistreated building, the spiral staircase, and entering into a room full of a golden light. Then the dream had continued and in that room her Maestra appeared and asked that she hand over the emerald ring that even she hadn't known that she had on.

'What importance can that ring have? Since childhood I've seen it in my dreams and in my visions,' she said to herself. The sound of a doorbell shook her out of her memories. The others of the group were arriving; they all entered at the same time and, by her count, there were twenty of them. Helena saw herself wrapped up in a fan of energies. All of the members of the group had very strong personalities. Most of them were young, but there were also people 50 to 60 years old. Her Maestra introduced her as the novice and this gave rise to various jokes. Helena was astonished to see that they clowned around a lot, but when it was time to study they changed completely, and became very focused and serious. The hours flew by and close to midnight they began to go home, saying that they needed to work the following day. Helena leaned out the window that overlooked the road and saw a street fight between the drivers of two buses. She said to herself, 'this place truly is horrendous for Marion.'

"I live with what humanity gives me and look what it is that they think I'm worth." Marion's voice took her out of her contemplations. Once more she had grasped precisely what Helena was thinking. "I could have all the money in the world if I would put my humble efforts into it. But I'm here for other reasons." Marion seated herself in her armchair and lit a cigarette. Helena shuddered at hearing this, and she remained standing, still listening to her. "For many years I've been protecting this place. I'm in this dump to prevent that the demons that they want to release get out." She shook her hands emphatically, signaling the surroundings. "The other day in the main square you saw what they do to activate them. When the Spaniards arrived they knew about these places and, together with the church, put seals to impede the escape of those Egos; that's why they built the churches on top of the ceremonial centers. When the 1985 earthquake happened those demons tried to escape; a few achieved it, but my presence in the site stopped the most powerful from escaping. If this hadn't been stopped, the catastrophe would have been worse; all of the center of the city and a large part of the country would have sunk. For a very long time I walked in the day and at night too, among garbage and rats, protecting the energy outlets to impede that these Egos would escape. I was alone, none of those that were with me accompanied me, but when it was time to go to the movies, they all appeared. Here, smoke a cigarette so that you're comfortable." She extended the pack to Helena.

"No, thank you." Helena knew that out of respect she shouldn't take anything from her.

"That's why I live here, with the help that Your Worships give me."

Helena felt ashamed to see that her Maestra, although she was helping all of humanity and struggling to deliver the Konocimiento, had to live in these deplorable conditions, deprived of comforts.

Marion began to speak once more. "One year before the earthquake of 1985, a disciple dreamed that he was walking in the center of the city when he saw that the buildings and the houses were toppled and there were huge cracks in the streets. In the place where he was there was an enormous hole, deep and black, in which the internal energy moved in a spiral towards the interior of the Earth and sucked in everything. This hole was joined to others that were smaller in size and all of them circled what were the ruins of the largest temple which they were excavating at that time. My disciple was terrified at perceiving these forces and at seeing the destruction that was everywhere. Distressed, he wanted to get to the house where I was living to see how I was, but there were lots of police that had closed off the area and they didn't let him approach. That was his dream.

One night before the earthquake," her Maestra continued, "the Konceptos indicated to me that I had to leave that very day. It was night when I took out my studies and belongings, just small things. The next morning, everything was flattened. There were a few people in the group who after a few days tried to recover my things from the apartment. A very large building had collapsed on top of the one where I lived. In the midst of all the rubble, my apartment was the only thing left standing."

In the following weeks Helena visited various locations. She went out with Marion and some of the people from the study group who were able to escape from their work commitments. The youngest were the ones that always found ways to arrive at these meetings. The persons who formed the group had very diverse interests and experiences. Some had been studying in the university or had an already-established career, or were office workers. Helena found an affinity with them and put a great deal of effort into practicing her Spanish so that they wouldn't poke so much fun at her accent.

One morning they went to the Museum of Anthropology where they went on a long tour. That day was full of strange occurrences. Upon entering the museum it seemed to her as though she was entering into another dimension; she felt like her head was numb and heavy. She perceived that many of the objects that were there were charged with many experiences, and it was this charge that was affecting her.

When they exited her Maestra made them walk through Chapultepec Park; it was four in the afternoon. They rambled through a wooded zone, forgetting that they were in the middle of an enormous city, until they arrived at the fountain of the frogs. It was an antique fountain, decorated with eight bronze frogs that sprayed water into its center. The retaining wall decorated with beautiful Spanish tiles served as seating for the small group. Of course, it wasn't long before someone began to play with the water, splashing his companions. After a period of relaxation they sat around Marion, who spoke to them of the motives of the karma of mankind.

"Man has a vital karma (his own) and a dosed-out karma, which falls to him to live due to the ethnic group to which he belongs and in which the same members symbolically establish their limits [...] Each group of Egos should find itself in the prescribed moment. Each group of Egos should find itself in a certain site and in a certain moment to fulfill its action (26 Mar. 88).

A specific moment arrives when *as law it is established that a group of Egos has to pass to a certain level and if they do not they*

fall behind and form a negative reason because they slow down those who are coming; when they do not pass they cause a great chaos due to the violation of the reason that prevails in that moment (26 Mar. 88). This occurs due to the development of the Fusion itself which is unfolded in certain HTimes."

This affirmation had left Helena dismayed. There is a particular moment in which it is marked that a specific Hueste (Host) has to fulfill its development, and if it doesn't, it affects all of humanity.

Her Maestra had repeated to her on various occasions that, "All of the events in the Fusion, and as a result, in the Planicio Terráqueo and among the Egos, are related; there are no separate reasons. However, we perceive things as if they were isolated events or acts, but in reality everything is connected. We perceive it that way because our konsciousness is poorly developed. This principle of interrelation is reflected in the Ego: if one Ego arrives at a specific comprehension, it evolves and opens the possibility that others do so as well; this, according to the level of konsciousness and awareness that each one has. In the same way, committing an error affects the rest. For this same reason, when a representative of a Hueste commits a fault or understands something, all of the Hueste, because of the sonorous relationship that exists, rises or falls in its level. When one understands, we can say that the konsciousness 'is opened.' On the contrary, when there is no understanding, the lesson needs to be repeated again. What we call karma is the repetition of lessons until they are surmounted. It is said that it's a step backwards because it's not permitted for new reasons to arrive until the previous ones are surmounted. It's as if we were in school; the end of the year arrives and those that don't pass need to repeat it. There is a karma for every Ego, another for each race, and another for humanity."

"So then the error began with the acts of just a few Egos in the past?" someone in the group asked.

"Yes, *the great errors and the great maladjustments are due to Egos that encouraged an error and founded what thou callest karmic reasons. Each Ego in matter is separated in itself, so that it, on its own, understands in an unparalleled eagerness that it not accept the connection to reasons that always lead it to an error. The mission of*

man is to understand himself; when he understands, the rest comes in addition. To comprehend is to be within his maximum. When man separates himself and distributes himself in a family, in a religion, or in a homeland, the karmic law is established. It is because of this that man should not have as the optimum, thoughts of bonding and generalized thoughts. Each man has a distinct intuition, and thus it should be, not linking his thought to a group or to another subject in particular, whether it be his family, or ethnic, patriotic or religious group" (13 Apr. 96).

She added that it was very important to understand that no Ego is equal to another; each one is different since the moment in which it was formed. For the same reason, there isn't any human being that is the same as another; what is essential for one isn't for the other. That's why it's contrary to evolution to try to link our thoughts to another Ego or to the structures that we have been obligated to adore. It's vital that each man be individual and that he guide his steps with the humility of surrendering himself to what corresponds to him. That's one of the reasons for which we were placed into matter: so that the Ego would find its imperative reason within itself, since each Ego has a unique purpose of Realization.

Marion explained about the Continuo (Continuum) so that they could understand why it is important to break with the structures that man himself has established. "The Continuo is all thinking action of man that envelops everything. This sentient action in reality is formed of sounds that reverberate and multiply themselves constantly, since we live in a Planicio of multiplication. It's been this way throughout the times and the ages.

It forms something like a densely tangled skein of thoughts that traps us until we forget that there are other reasons that exist outside of it. And we end up believing that this is reality. The 'entanglement' or Continuo (Continuum) absorbs all of man's thoughts since he is born. This becomes like a magnet that pulls in all of mankind's thoughts. The majority of these thoughts are pointless reasons but they have great force because they have gone on reaffirming themselves throughout time. We will only be able to step out of this Continuo when we separate ourselves from human reasons and interests. We

need to be individuals and stop being influenced by what all of the world wants or thinks, because this is counterproductive to evolution. We can achieve this with comprehension, in understanding who we truly are, and how we are related to what lies beyond this fantasy. With comprehension (and study of the Konocimiento), *each one of us will project a 'purer' sound, less contaminated by the Continuo (Continuum), until we are able to couple ourselves to the Kosmic reasons to which we belong, leaving behind the karmic reasons that currently give strength to the Continuo"* (Nov. 10).

"Why do we always repeat the same mistake?" someone else asked.

"Man has a great disharmony, and he has not wanted to adjust his interest to the development of the Universe. *Each particle* in the Fusion (like the Ego) *embarks upon the task of its participation by means of one of the many compounds of Konciencia (Konsciousness) and it adjusts them to its interest and to its responsibility. In the case of man, it adjusts them according to his interest and not to the responsibility of his acts in relation to the primary manifestation* (the Konceptos). *This attitude has been reaffirmed throughout the HTimes. Today we can say in truth that the other humanities had a direction that was very different from that of the present humanity, but primarily there was always within them that of entertaining themselves in their own interests"* (17 Sep. 99).

"What have those interests been?" they asked Marion.

"All that which had led man to enhance his image as an all-powerful being who can poke around here and there without caring what he brings about. *Man has interior regulations that should lead him towards a truth. When he didn't accept them, he became confused in his way of thinking, and that is why, and throughout many remote ages, he took shelter in a respect towards an ethnic group, and later, towards a homeland. This is in compensation for his initial reason, for which what thou callest religion was established, formed of images that shaped a parody of that which in truth should have been understood. To distract him he was given the responsibility of a wife and children. All of these conceptions were made by the lords of the darkness, and have been perfected in the course of the HTimes, up to what today is*

123

manifested in completeness (17 Sep. 99). These reasons are totally contrary, since man should revolve around the Kosmic reasons, which are the truth within him. Now man believes that he only lives one lifetime and as a result puts all of his interest in material achievements. He lives in a world where he only wants to have a good time, surrounded by comforts, dreaming that illumination and the favor of a god will fall from heaven and that one fine day he'll wake up knowing everything, or he dreams of a future in which man takes shape as the master and lord of the Universe. He thinks that the goal of life is to find happiness and looks for it in a family, in a job, or in a doctrine. But this is a deceptive quest; it's like trying to trap water between your hands and it only brings him anguish and lack of satisfaction that absorb his energy 24 hours a day, because he's searching for something that isn't real. Man will find satisfaction when he understands who he is and develops himself accordingly."

That day Marion and the group had chatted for a long time about the changes that would come with the move from the city and of the difficulties and tests that each one would face upon leaving everything

behind. What might seem easy to one person was something very complicated for another. But each of them accepted it and made the decision to continue.

"Now that each of you has decided, you'll know what you should do," Marion said to them.

They noticed that the sunset had begun to color the sky dramatically in contrasts of reds and greys; the temperature was very pleasant. They picked up their bags and started walking towards Reforma Avenue. They had returned to their games and their jokes. This was something that amazed Helena: the readiness that they had to laugh most of the time.

They accompanied Marion to her house and said goodbye, agreeing to meet the next afternoon to continue with the plans for the move. Helena walked to the hotel, reflecting on how Marion managed so naturally in Mexico; she knew the places well and spoke Spanish without trace of an accent. It seemed that she had lived there all of her life. But the same thing happened in Paris. She was bewildered. Where could she be from?

The next morning Helena awoke anxiously; what she had seen in her dream really frightened her. She felt the need to talk to someone, but she'd have to wait until the afternoon.

When they were all gathered together, Helena began to relate her dream and her voice trembled with the uneasiness that the memory of the images provoked.

"I see myself in a place that's very dark; even though my eyes are open, I can't see well. I hear and I feel that something is falling from the sky and I think it's rain. I start to walk, but I can't; my legs sink and keep sinking up to my knees. When I pull out my leg to take the next step, I see that it's clean, it's not wet or dirty; rather it feels warm and pleasant. I realize that it's not mud like I thought at first and I see that neither is it raining. There are many noises, as if they're coming from the depths of the Earth. The ground moves and when I look up I see some mountains in the distance, and there are lots of lights coming out of them, like fireworks. As I watch them I realize that I'm flying and I can see what's happening all around; I see trees and undergrowth. From the mountains come lights, smoke and a kind of mist that little by little takes the shape of phantasmagorical faces and bodies. I woke up very unnerved because of what I felt and heard." 125

She finished telling them the dream, feeling anxious all the while.

They listened to her, but didn't give the dream much importance. They were so busy planning the move that they placed all of their interest in that and forgot to study the dream. The study that afternoon was to decide to where they were going to move; they agreed to search for a place in the south of Mexico, in the states of Chiapas and Tabasco.

When they had all left her Maestra's house, Helena remained behind to wash the coffee cups that had been used.

"Can you imagine a group of spoiled city-dwellers arriving suddenly at the jungle? Honestly, I don't know what they're thinking. Living in the jungle is going to be extremely complicated; it's very hot and the environment is really rough. Most of them have spent their lives in the comfort of a city and when they go on vacation they just go to a hotel. They haven't thought that they're not going to have water or electricity. And not to mention what they're going to live on," commented Marion, worried by what she saw coming. "Well, it's all going to be part of the tests that they'll need to pass so that the Konocimiento is delivered." She added, more for herself than for Helena, "The hardest part is going to be the battle with themselves." Suddenly Marion asked her, "And you, now that you've decided to come here, do you already know what you're going to do? Or are you expecting that they serve you at the table?"

Her words shook her deeply and she wanted to argue: she had barely arrived and didn't know what the situation was. She felt a warning in her body; she knew that she shouldn't give in to her irritations. Helena realized that she really hadn't foreseen anything. She felt uncomfortable with herself; a thought of insecurity crossed her mind, 'maybe I don't have the clarity that's necessary for this path. Maybe that's why Marion left me far away in France while she was here with the rest.' She felt confused and alone.

From this moment on they began to organize themselves in groups of three to five people who took turns going to explore the area. They made several trips and visited different places in Tabasco and Chiapas. Most of the towns that they saw were settlements where they raised cattle and others were places that were very swampy. The cities weren't what they needed for their interests.

After several attempts to find the right place, five people from the group decided to go to Palenque to investigate the area and try to find a plot of land that would serve their plans. They left on the bus for the city of Villahermosa, where they decided to spend the night. The next morning they took the bus towards Palenque and arrived hours later. It wasn't very remote, but the transportation was quite poor.

Palenque was a small town that was close to the Mayan ruins. There they could get only the basic necessities. There was a small plaza in which a market was set up and everyone went there to buy their provisions. But what most amazed them was the movie theater: it was an open space, without a roof, with a large wall in front where they could project the films. The seats were wooden slabs placed on bricks. Exhausted by the long journey and the tremendous heat, they found a tiny hotel with four old rooms where they spent the night listening to the squeaking of the rusty fans that hung from the ceiling.

That first night, one of the five people dreamed that they would find a parcel of land on the highway heading towards the pyramids. Everything began to fall amusingly into place from this moment on. The following morning they set out enthusiastically to investigate. They walked for a good while along the highway and were totally drenched in sweat, as the humidity of the place was overwhelming, but they didn't feel tired; the contact with nature stimulated them. They were amazed at the quantity of sounds, colors, and smells that inundated the landscape. It was a completely uninhabited zone, not a single car passed in the entire time that they were walking. After two or three hours of exploring the location, to their surprise, the woman who had dreamed of the piece of land recognized it.

It wasn't possible to see it well because of the amount of undergrowth that covered it; it was in the middle of the jungle, even though they were just a few kilometers from the town. They didn't dare to walk around much in it, since they were only wearing sandals and they thought that there might be snakes in there. Their caution was justified, as several weeks later they found out that the name of the place meant "land of the *nauyacas*," which is one of the most poisonous snakes in the Americas. The site was flat and bordered a stream of clean, clear water, where small fish could be seen swimming. The jungle was enigmatic, full of possibilities and mysteries. The songs of the birds and the noises of the insects filled the air. Excited, they went to find out who was the owner of the land.

Remembering that moment months later, they had laughed upon seeing that all had been guided by the Konceptos. Just when they set out on the highway to return to town, they bumped into the owner. The man seemed quite taken aback to see this very odd group leaving his piece of land; he didn't understand what city people were doing there. They started to talk with him and they asked him if he would allow them to live there, at which, he, fairly astonished by the petition, responded yes; they'd just need to cut the underbrush and clean up the land in exchange. They returned to town and called Marion to tell her the good news.

The days that followed were full of commotion, and those who had stayed in Mexico City busied themselves with the necessary preparations for the move to the jungle. No one had lived in such a place and Marion helped them to understand what would be useful to take along. In addition, they had to be careful of the scarce amount of cash that they had managed to scrape together; some put in their savings, others the compensation that they had received upon quitting their jobs and the rest was the little money that Marion had. One of them traded his car for a truck that could transport all they would be taking. They bought the basics they were going to need, foreseeing that in the little village they wouldn't be able to find almost anything. They bought cots, mosquito nets, and sacks of rice, beans and lentils. Marion sent Helena to a canvas manufacturer to make a tent so large that the man thought that they were requesting the big-top for a circus. That tent would serve as a large dormitory. Helena visited him daily to see if the job was being carried out well and she added the small modifications as they arose. When it was finished, the moment of departure had arrived.

After five weeks of preparation, they left for Palenque. When they got to the site, they found the land to be completely wild. The five who had arrived previously had not been able to clear it while they were waiting for the rest of them to get there. The weeds grew so rapidly that when they had cleared one part, they grew back again in a few days. Furthermore, things had become complicated because upon clearing the plot they found *nauyacas*, scorpions, spiders and all kinds of insects.

With the arrival of Marion to the site, the projects started to be carried out little by little. She organized them and after much struggle they were able to clear the land; they began to build the septic tank and they built benches and a table from tree trunks that they found on the same spot. They got some large beams that served as support to erect the tent, and finally, after much effort, were able to raise it. They then built a palm-leaf shelter near the stream that would serve as a dining hall and a place of study. They prepared a burner in order to cook food. Adapting to all of this hadn't been easy for anyone. They had no comforts. There was no water or electricity, and the heat and humidity had affected the health of some of them.

Life in the jungle was astonishing; the place had an impressive beauty and the different tones of green shone with an intensity that stunned the senses, just like the sounds that came from the huge variety of animals and birds. At night the jungle reverberated in a cacophony that at first prevented them from sleeping, but they became accustomed to it to such a degree that when the nights were quiet, they slept uneasily.

There was an abundance of armadillos, raccoons, iguanas and coatimundis in the site. These animals approached them with curiosity. They observed them as if they wanted to ask, "And you, what type of animal are you?" Helena got close enough so that she was able to touch them, and they didn't run; they were attentive, watching her. The iguanas and the birds were the most receptive to contact. When the group became accustomed to seeing snakes, scorpions and other pests, they realized that these didn't attack; they just kept on in their path. Helena was surprised to see that the butterflies flew in groups according to their colors; red with red, yellow with yellow, the spotted with the spotted. It seemed that they followed a well-planned choreography; they flew together but never mixed between themselves, even when all of them settled on the plants. Everything in the environment lived in agreement, harmonized by the energy that emanated from the earth, powerful and primordial.

The heat and humidity were intense and constant, above all after the rains that fell in abundance during the summer. These were strong torrential downpours that lasted several days. For Helena they were a gift; she became accustomed to working and walking in them, enjoying the moments of respite from the suffocating heat.

The enormous rubber and *ceiba* trees surprised Helena with their size and lushness. During the day as well as at night, the place emanated a very potent energy that made her feel more alive than ever before. The sky studded with stars softly illuminated the silhouettes of the trees and that nocturnal brilliance allowed them to walk through the jungle. This vital coexistence helped her and the others to bear the lack of comforts.

Even though their diet was quite limited, since they only had beans, rice and lentils to eat every day, the energy of the jungle gave them sufficient vigor to be able to withstand the heavy workload on the plot of land. In addition, the long hours of study nourished them. Helena observed how Marion always bore the extreme conditions with good humor, even when her companions began to fight amongst themselves over unimportant things.

In the beginning they depended on the stream for bathing and for drinking water. They had to make several trips each day carrying the heavy buckets; it was a task that irritated and tired them, when before in the comfort of their homes, they only needed to open the tap to get water. In a stroke of good luck, one day they found a small house close to the lot that, incredibly, they hadn't seen before due to the quantity of undergrowth that covered it. The house had a well and the owner allowed them to connect a very long hose that crossed his property so that they could have access to the water for drinking or bathing whenever they wanted. It was a luxury that they were all grateful for.

Daily life was a struggle, and as they had very little money to buy the necessities, they had to find an economic solution. Before leaving the city, Marion had foreseen that it was going to be difficult to make money in a place that was so remote and with so few possibilities. One day she had gone walking in the center of the great city and found used clothing, lamps and knives which they ended up selling in the flea market in the village of Palenque, with the items placed on a rug on the ground. They didn't earn much, but it was enough to buy the basics.

And so the months went by. In the daytime they fixed up their home and in the afternoons they met together around Marion to study. The small community began to form itself around the studies. Her Maestra spoke to them about the motives of the Ego and of the different HTimes and teachings through which it had passed, of how and why man had entered into the reasons of karma and of the consequences of the abuse of Konocimiento.

One night Helena dreamed of some strange images. In her dream she saw herself walking along a path in the jungle. The road was almost imperceptible through the profusion of foliage and she had a hard time walking because of it. At the end of the path she found a cave. When she entered it, she was surprised to see some large stone tablets carved with Mayan figures. She told Marion about it.

"I understood that I was seeing slabs of stone that were carved here in this place."

Marion confirmed that it was so. That afternoon, she talked to them about the meaning of Helena's dream.

"Our stay in this place is going to help us confirm what we have studied up to this point and what we will continue to develop. Later on we'll find some Mayan stone slabs that speak of the history of mankind. This information will give the world a different vision from that which historians and scientists have presented to us. This will help us complement, and above all confirm the studies that we have. What the Mayas had was important for their time, and what we have now is what follows and it surpasses it because we're in a different time. We'll stay here for a while, but then we'll go to another place deeper within," she said as she indicated the jungle, "where we'll find these stones. If someone finds them before us, they may even say that the images on the slabs are of extraterrestrials that visited the races of these lands. But the carvings are only representations of the demonic Egos that escaped before their time and imposed themselves here as their gods. We know that the Earth is isolated, and that it isn't permitted that prohibited reasons either enter or leave. They speak of extraterrestrials to cover the truth: that these are Egos in rebellion that in the past as well as in the present reject the limitations they have to live."

Since they had arrived at Palenque, Marion constantly searched for ways to encourage them, to gather them together. Her mind was always full of ideas and projects. Apparently, everything seemed to be fine; they got along cordially and they strived to get ahead in the material reasons.

After almost a year in the place, small changes in the environment began to occur; there were small, isolated earthquakes and rockslides, a crack opened in the earth near the encampment, and sometimes as fine powder had fallen. The temperature of the water in the river varied; sometimes it was hot and smelled of sulfur. The heat was more intense than usual. They went to town to see if someone could explain these changes to them. They only said that a group of foreigners had been doing experiments in the area, but no one knew what they were for and neither did they pay them much attention.

The weeks passed until one night, almost at daybreak, Helena awoke upon feeling a slight earthquake, but what most impressed her was hearing that the jungle was silent. Then she heard an almost imperceptible noise, as if something was falling over the tent. Thinking that it was rain, she put out her hand to check, but when she pulled it back in, she saw that it was covered in a warm powder that was as fine as talc. When she saw that the others were still sleeping, she went out to see what was happening, but it was completely dark; she looked at the sky and didn't see a single star. She noted a strong odor of sulfur and there was something in the air that didn't allow her to breathe well. She covered her mouth with a towel and left the tent to go look for Marion; in that instant she heard her voice calling everyone.

Marion walked out of her house with a lit flashlight and together they went to awaken the others. Her Maestra yelled at them several times, "Wake up!", but they didn't pay attention; they all started to grumble, saying that they were tired and wanted to sleep.

The powder rapidly covered the roof of the tent and it was on the brink of collapsing under the weight. With broomsticks Marion and Helena began to push the roof upwards to remove the powder that had fallen. Finally, between grunts and curses, the others awoke and began to help. When they all walked out and saw the darkness that enveloped them, they became frightened; they didn't understand what was happening, and the earthquakes, although gentle, were becoming more frequent. The reactions of each one of them were very different. Some laughed, others cried and some screamed, afraid that they were going to die.

Suddenly they heard an intense sound, as if something had ripped within the Earth. Because of the large echo that exists in the jungle, they heard the screams and lamentations of other people as if they were nearby, but in reality they were quite far away. The heat and the stench of sulfur became unsupportable. The sky began to illuminate itself with lights of many colors. They heard that the current in the river was very strong and went to investigate what was happening. When Helena tried to walk, she realized that she was buried in the powder up to her knees. The memory of her dream came back to her in that instant. She looked up at the sky and saw it lit with sparks of colors. They were able to reach the river and saw that it was swollen and had swept tree trunks, fish and animals along in its path. The water was so hot that the fish jumped out of it because they were burning.

The hours passed and everything continued in darkness, with the exception of those moments when they saw great blasts of fire coming from the same spot where they had seen the lights in the sky. The earthquakes were continuous and they ended up getting used to them. They went on like this for an entire day, and then another, waiting for the powder to stop falling. They took turns removing it from the roof.

At dawn on the third day, Marion was angry and desperate because of the lack of light. Looking up at the sky, she raised her right hand to emanate and signaled with her index finger towards a point in the firmament; she asked with all of her might that the sky open and that the clouds of powder leave. The force that was projected from the hand of her Maestra was incredible. Helena perceived a ray of lux (light) shoot out of her emanation. In this very instant a circle began to open within the clouds of ash. It was a small point that grew larger and larger until it allowed the rays of the sun to enter and finally they could see clearly. Helena was speechless; she was stunned by what she had just seen. She realized how little she understood about who her Maestra was. Helena thought that Marion was much greater than she had realized. Who could she be if she was able to open the clouds and emanate rays of light? If she hadn't seen it herself, she wouldn't have thought it possible.

When they were able to see their surroundings, they were speechless. Helena felt a knot in her throat. The jungle, the exuberant, dense jungle brimming with life was now an ashen wasteland.

Gigantic trees like the *ceibas* had fallen because of the weight of the dust; the vegetation was covered with it and burned, the animals that had survived ran frightened without knowing where to go. Huge quantities of birds, insects and animals died because they couldn't breathe in the thick ash that had fallen. The water in the river looked like a mass of mud; it smelled of sulfur and there was no longer any life in it.

They wanted to go to the village, but were unable to move the truck because of the depth of the ash layer. They had to walk along the highway to the town, but their progress was difficult since they sunk in up to their knees, as though they were walking in snow. There was no means of transport. Further on they found some men that were removing the powder with a backhoe. They told them that a volcano had erupted, ejecting enormous quantities of ash and that the clouds of dust already covered various states of the country. Afterwards they found out that the cloud of ash had gone all the way around the world several times.

No one, not even the peasants that had lived in the area all of their lives, had thought that the volcano could become active. The destruction in nearby locations had been terrible. Entire towns had been buried and burned under the ash. The explosion had covered an area 100 kilometers in diameter. They heard many stories relating the misfortunes that the eruption had caused. Helena was surprised to see how, in spite of the problems they had, the group hadn't really been affected, in spite of their closeness to the volcano. In that moment she knew that they had been protected and remembered the way in which her Maestra had parted the clouds of ash. She felt a great love and respect for all that she represented, even though deep down she felt a slight fear of being before the greatest presence that man could ever imagine.

In the days that followed it was very difficult to try to find water and food, which were scarce. The little shop in the village had been emptied immediately. In the middle of this grey place and with all of the problems of survival that they needed to resolve, the duality of centuries manifested itself in the group. The fears began to show themselves in anger, arguments, and in the lack of clarity. Marion tried to gather them together, saying that this time was a great test for all, that the important thing was to continue ahead, and that if they were able to remain united they could overcome the obstacles. But they no longer understood why they were there and they forgot what had motivated them to go and live in the jungle. They only wanted to return to the comforts that they had known before.

The highway had barely been opened to traffic again when two people left the group. They were frightened and said that this wasn't for them. Marion explained to all of them on several occasions that the tests they needed to face came at specific moments in which the Ego was pressed in its deficiencies or weak points so that it could confront them and thus, resolve them. Helena, in her fears and confusions, tried to cling to what she had learned and to all of the experiences she had shared with her Maestra.

After a long month of hardship, the much longed-for rain arrived; strong downpours fell and cleaned everything, the jungle recovered its color, that brilliant green that had so impressed Helena when she 135 had recently arrived. But the longing they had all had at the beginning just wasn't in them anymore and little by little they began to leave. Helena hadn't understood how it was possible they could leave after so many experiences, studies and labors that they had carried out. At the end, they had only mumbled to themselves that, "I'm going because I was just passing the time and now I want to do something different." Helena had become angry upon hearing these words. She couldn't understand how they could reject all of the struggles of her Maestra and the Konocimiento that she had given them.

Helena thought of the dream that she had had before coming to the jungle and realized that importance of the scenes that she hadn't understood at that point. She remembered the figures she had seen coming out of the mountain in the dream and she asked Marion, "Do you think that those figures were the Egos that you told me about in the main square in Mexico City that night? The ones that come out of the *bocas* (outlets of energy)? We were here to prevent those demons from escaping, right? Do you think we accomplished it?"

Marion looked at her and Helena saw an immense weariness in her face. She responded, "It wasn't a total defeat, but the energy that was arriving here could have helped us to progress. Great advances were going to come for mankind, but they didn't appreciate it, they preferred to continue on in their same sad stories as always." With that, she walked slowly towards her hut.

She felt her Maestra's pain; it was beyond sadness and disappointment. It had been many years of having lived with them and of having put all of her effort into each one. She had imagined them strong, secure and battle-hardened, teaching and protecting the Konocimiento, but the ashes of a volcano had finished off all of their desire to evolve. Helena and Marion were alone. The idea for which they had moved to this place was no longer valid. They took care of the few unresolved matters that remained and returned to Mexico City.

Helena studied a couple of years more in Mexico at the side of her Maestra, going deeper into new roads of Konocimiento.

"If one door closes, another has to open," her Maestra said. "You'll see that new opportunities will arrive." At her side, Helena continued perceiving more and understanding that which she had told her so many times. Sometimes she was frustrated by her slow progress or by her failures. Marion explained to her that in order to arrive at a real understanding a process had to be lived several times. "It's a slow process; we're like an onion, covered in layers and layers of veils that don't allow us to see clearly. One day you think that you understood something and later on the lesson returns to you in another form and you need to confront it again; it continues on until it really is overcome. This is a process through which the Ego will have to pass in order to adjust itself completely. There will come a moment in which it perceives with clarity its reasons on the Earth in order to incorporate itself once again into the Universe and continue on its path."

One day Marion called her so that she'd go over to her house to see her. When she arrived, she found her seated at the dining table; she was working on a bonsai that a neighbor had given her. It was a pine with a twisted trunk. With great care, Marion was removing the wires and nails that bound its roots and stems and impeded its growth. She spent an hour working in silence with all of her attention focused on it. Helena observed the gentleness of her movements and the care she took to not damage the tree. When she finished she moved it to larger temporary pot and said to Helena,

"Now it doesn't have shackles that hinder its growth." Helena helped her clean it and pick up what she had used. Marion added, "That's all I've wanted to teach all of you, that you break the restraints that maintain you anchored and limited. The man who is free of his deficiencies can develop himself and project himself to other places of greater plenitude, of course, without altering the established order." Helena listened to her, moved. "I only came to give all of you the tools so that you can achieve it; the Konocimiento is the only way to get there."

Marion took Helena gently by the arm and they sat in the living room. "There's a reason for continuing in Mexico, but for the time being you have other things to do. Go back to France, there are specific memories there that you need to have. You need to understand what happened to the teachings that were delivered in the Second HTime so that the same thing doesn't occur again. You have the greatest gift a human being can have, the Kommunication with the Supreme Entities that form us. You are on a pedestal of glory, but it's your duty to understand it. You must take care of the Kommunication that you have and you must make whoever crosses your path with a sincere intention, understand it."

Marion gave her a portfolio and an envelope. She said, "I deliver these manuscripts to you. They will be the foundation of the studies to come. People will come to you who will cherish the Konocimiento and in this way it can reach the whole of humanity. Remember this, if one arrives, all arrive. If one understands, he opens the way for the rest to understand."

She went to her room and took out a small red box. "Look, I have something for you; I've kept it for a very long time." Helena took it in her hands without knowing what to do. "But open it, Helena. Or don't you like surprises?"

Helena opened the box and found a ring inside. It was very ancient, of gold, and had a round emerald mounted inside a thin circle. The band was engraved with geometric designs. In this moment she remembered the vision she had had when she was a girl. She couldn't believe it. "But how is it possible?" Her cry seemed to fill the entire room.

"Don't look at me, I don't know, I only had to give it to you." Marion looked at her, her eyes twinkled and she urged her to put it on. Helena tried it on; it was a little big for her ring finger, so she put it on her index finger. "Umm, it's a little big. Well, in time it'll fit you."

Helena read between the lines. "But, it's…" She couldn't continue; many ideas passed through her mind. She looked at her Maestra, waiting for her to say something else.

"When you return to France and remember, you'll be the one that tells me about it and I'll tell you how it came to me. Have you seen how it's set, the stone has a tremendous brilliance and if you see that it's really clean, it's because I was polishing it and…" Helena didn't hear any more, she was thinking about the meaning of the gift.

When she arrived at her apartment that night, Helena opened the envelope and read:

> '*It is said that great changes are coming for the humanity of the future. [...]*After an era of intense cold, will come the thaw. [...] *Due to the new heat, intense, the reptiles will fly. The animals in general will have great mutations and will be completely distinct from the current ones. For example, the insects will be defeated by the mental power of men, and animals like the horse will have the power of levitation, these having moreover, horns as in the past.*
>
> *There will come for us eras of great disputes, half of the world against the other half. The Lux will confront other beliefs, when all of these perceive different ideas[...].*

In the future sentiment will not prevail as it does in the present, rather it will be the use of the energy projected in all that is undertaken. There will not be emotional relationships, but rather one will understand and one will live knowing that everything is a power, a force, an energy. This same energy will project itself, for example, in the construction of floating cities, in which by means of levitation one has the power to move what one wishes with a finger, and where one lives momentarily and in harmony, so to thus inhabit a Planicio Terráqueo that is like an Eden, where nature is respected. As well, man will know how to use the water that is found in the air, and thus the water will never run out.

Another form of energy will be projected by means of the mastery of the mind, in order to propel a nobleness in the Ego. The written language, as much in letters as in numbers, will suffer transformations until it becomes "symbolic," like that of the Mayas, in which we see a compressed word, but whose sound will be an open door so that each pequeño vitalizes this word with his own elevation, and perceives in this way a great many reasons by means of the relationships that he can establish between the different sounds; it will be like an empty house that each pequeño will decorate and will make functional according to his interests. This will be a preparation so that the pequeño has telepathy in completeness and doesn't have to turn to his matter for a profound expression. Thus each Ego that reads a text, will indeed be the writer himself who vitalizes, adorns, and realizes in truth the text by means of his perception; and the Ego who writes or expresses a text, will be the reader of what is presented to him in the midst of a given moment and he expresses it like a mother idea, that the rest develop according to their capabilities.

The word will be like a whisper in which the force will prevail as a vital characteristic, and it will be the modulation of the sound which provokes therefore an ample projection in the mind of each Ego.

When man speaks it is as if he crawls. Upon transmitting a thought by telepathy it is as if he were flying. Thus, the spoken word, in the future will be like writing, and this new "symbolic" writing will be like the spoken language of the present. This will make everything take its own level and the necessary differences are established for harmony: fleas with fleas, lice with lice, etcetera (23 Jul. 92).

The Egos will be divided according to their level of konsciousness so that the evolved ones are not affected by the negative thoughts of the less evolved. Negativity is only a lack of understanding.

Helena, you have recognized some of the errors that were committed in the past. But one doesn't always walk through dark tunnels. This will give you a taste of the magnificent future that awaits humanity.'

Chapter 8
The Sequence

One who is said to be "wise," but is half crazy,
the same as his son who was an idiot,
says to a powerful cripple,
"Let's conquer the world with a bomb."

REPERCUSSION

The light that illuminated the city of Paris at that hour of the afternoon always inspired Max. Even after so many years of living in the city it continued to amaze him, but lately he was indifferent to it, ever since his life had been turned upside down. For months he had been in a constant state of agitation. He missed the days of his youth when everything seemed simple to him and full of possibilities. He asked himself, 'what happened to the curiosity and openness that I once felt?'

He went out to walk along the avenues of the capital to see if it would help calm him a bit before his meeting with Helena. Even though he sought this encounter, he was anxious because he didn't know how she was going to react to what he wanted to tell her. 'So many years have gone by since I last saw her,' he said to himself. 'What will she be like now? Will I be able to talk to her like I used to? I hope that she can help me to climb out of this hole.'

Max had thought of Helena many times throughout the years. He remembered her as a brilliant student, with a great intuition, restless and questioning the reasons behind everything. He didn't know what had become of her since the last time he had seen her. He did a quick count; almost 14 years had passed since then. She had left the University suddenly in the spring of 1991. Then, during the following three or four years, he didn't recollect exactly, he bumped into her a couple of times on the streets of Paris. In those encounters she chatted with him very enthusiastically about some of the things she was studying. Although Max didn't understand everything she said, the things that she spoke of were very interesting and unknown to him, and for some reason her words remained in his memory with an unusual clarity.

He remembered her with fondness: the expression of her eyes, attentive to all that occurred around her, her mischievous smile and that unusual mole on her forehead that gave an air of mystery to her lovely face. He was surprised that he recalled so much about her, but above all, her peculiar manner of seeing things, even though sometimes she seemed to go too far in what she proposed.

He set himself to thinking about what he wanted to tell her. A year had already passed since he received the news that completely changed his world. His younger sister, Isabel, had been diagnosed with leukemia and Max took charge of her treatment. But in spite of all of his efforts to get her the best doctors and the newest treatments, she continued to worsen. Witnessing the process of her decline was painful and hard to accept as her brother, and more still as a scientist. In the last months of her illness she had asked him to let her die in peace; she didn't want to receive any more treatments. Three months ago she died with a peacefulness and acceptance that Max hadn't understood. He grieved for her loss, but worse than the sadness that he felt was the frustration and anger that came from not having been able to save her. He dedicated himself to cancer research, but he didn't have the power to cure his own sister. The absence of Isabel left him submerged in a great sorrow. They had lost their parents when Max was 19 and he took charge of her, as she was five years younger than he. They were extremely close because of all they had shared. Now Max couldn't concentrate on his work and on top of it all, for the last four months he had been dreaming of strange images that he was unable to understand. The dreams repeated themselves constantly, distressing him; he couldn't find peace either during the day or at night.

During these difficult months he thought many times about Helena. He remembered that she had spoken to him often of her dreams and that for her, they were an important part of life. He saw her as a possibility for helping him find the answer that would allow him to be at peace, but he didn't know where to find her. Years earlier he heard that she was living in Mexico; that was almost 12 years ago. But as destiny is unavoidable, a couple of weeks ago Max was walking along the streets, trying to calm his mind and he stopped at a magazine stand to look at an issue that caught his attention. On the cover he saw a photo of an old man with his grey hair pulled back. The photographer had captured in the expression of his eyes, the experiences through which the man had passed. The portrait had an unusual depth. Curious, he opened the magazine and read the photo credits, seeing the name of Helena LeMond. 'Wow, I'm in luck!' he thought, congratulating himself. Buoyant, he headed to the publisher's office looking for a way to get in touch with her. And so it was that two days ago Max received a phone call from Helena, saying that she'd see him today at 5 in the afternoon.

His thoughts returned to the present while he continued walking along the Rue de la Paix. He stopped in front of an antiques store to look at the watches that were displayed inside. He was surprised when he saw his reflection in the window. He noted the bags outlined under his blue eyes and the weariness registered in his face and sighed. He was a man at the middle of his life, still strong, but his posture was now a bit fatigued with the passing of the years. He ran his long, thin hands through his wavy hair. The grey hairs were more noticeable within the profusion of the black ones which were fewer every time he looked. He saw his exhaustion reflected, but he didn't want to pay attention to it and continued on his way. The scant rays of the sun reached his back and he felt its gentle light and warmth on his neck. Surprised, he turned and saw how these last rays illuminated the silhouettes of the buildings, announcing the arrival of dusk. He had to hurry if he was going to get there on time.

Seated in the cafe, the hours had flown by for Helena; the same thing always happened when she studied. It was as if the time unfolded itself and she entered into a different awareness. She had been studying about Max and understood that his world was collapsing on itself so that he could break away from the habits he had repeated for centuries. But he would have to realize why he was arriving at this point.

She checked the hour, closed her notebook and capped the dark green fountain pen. She enjoyed writing with it when she studied because it marked the difference between the everyday words and those that she received in the moments of study. Marion had made fun of her saying, "Ah, you took out your magic pen; let's see, what's it going to tell you today?" Helena touched the small mole on her forehead lightly and looked around; the people continued walking from one side to another, most of them in a hurry, as if something important were going to occur.

'How many faces, how many lives…,' she thought, while she watched the people pass. It was in the ones who walked alone that it was possible to see their thoughts more clearly, since they weren't so worried about posing. She thought of how, throughout the ages, people had saturated themselves with habits, with masks, with disguises in order to continue participating together in what they considered life to be. She thought of Diego, feeling his uneasiness at confronting his father's illness. 'How different would it be if each person knew that this life isn't the only one, that death is simply a transformation and *that his life, anxieties and problems will only last a second. That he does not die, that he lives and will continue living in evolution during Seven HTimes and beyond that. […]He will realize that his small mistakes like wars, catastrophes and the rest are not important so that he doesn't persist in them*' (4 Nov. 07).

A cold wind began to blow among the bare trees that bordered the park. Helena gathered her chestnut hair in order to adjust the new scarf that she wore around her neck. She had taken advantage of her trip to Paris to visit the clothing stores. She loved to go shopping and every chance she was able to indulge in something, she did. When she had seen the fuchsia-colored scarf in the shop window, she couldn't resist: the color reminded her of the flowers of Mexico. She liked to make a contrast with the grey tones that dominated in Paris in this season. She breathed deeply looking at the clouds that began to form themselves in the wintery February sky. She lit a cigarette, inhaled the smoke, and called the young waiter over to indicate that she wanted more coffee. These two aromas always produced a pleasant sensation and intensified her concentration.

Helena recalled the years in medical school while she awaited the arrival of her old classmate. She had felt an affinity with Max when they studied together and they used to run around the city with a small group of friends. Almost 14 years had passed since the last time she saw him. She remembered him as a young man of 28, full of curiosity and energy, very focused, and with the determination that one day he would discover something important. He had always been restless, with a great urge to learn. He had been fascinated by the complexity of all of the systems that make up the human body. Nevertheless, Max was an arrogant student, desirous of leaving his name recorded among the great scientists that would lead mankind to a better life. For him, the body was a mystery he wanted to understand in order to "improve it." But she also remembered the other part of him, that which glimpsed that there was something that existed beyond reasoning and experimental quantification. He was very intuitive, but he was afraid to recognize it and he hid it. They were two extremes that he confronted within himself constantly. Helena perceived that his call had to do with that internal battle. In the past, Helena chatted with him on various occasions about what she was learning in the first months of her studies with Marion. Max had wanted to know more about those reasons as long as they didn't rattle his tangible world too much. He had had a special interest in genetics and Helena asked herself now if he had found what he sought, and even more, if he was konscious of what motivated his quest.

Diego walked out of Doctor Manet's office, in a hurry to get to his appointment with Helena on time. He had visited Paris rarely and the size of the city always left him disoriented. He got into a taxi to go meet with her in one of the cafes in the garden of the Palais-Royal, thinking about how everything had occurred unexpectedly. He took advantage of the interval to remember all he had lived in the last few days, trying to find an equilibrium within the whirlwind in which he was involved. Some days earlier he asked that he would find clarity to be able to understand how to resolve the situation with his father. And here he was, talking to a specialist and about to see Helena.

Diego had been studying formally with her for two months, being initiated in the studies. He had asked her to teach him; the things that she instructed him permitted him to "assemble" a new way of seeing life, as he said to her. The idea of being able to develop a relationship with the Supreme Entities and Konceptos that sponsor the action of the Ego seemed marvelous to him. Helena had commented to him that the openness he had in these moments of starting out on his path was invaluable in order to be able to relate to them and thus begin to know himself and the environment. She told him that many tests would come to him so that he would arrive at understanding himself more fully.

In the last six or seven weeks the condition of his father Esteban had worsened rapidly. Diego was overwhelmed, thinking about how he was going to resolve the situation. A week ago the doctor in Avignon told him that very little time remained for his father if he didn't submit to a drastic solution.

Diego felt odd, unable to explain clearly what it was that bothered him about knowing his father's death was imminent. It was in these somber moments that he received a phone call from Helena asking him to go to Paris, without explaining why. She just said, "I'll expect you on Thursday in the cafe in the garden of the Palais-Royal at 5:30 in the afternoon." At first, Diego had felt irritated because she was asking him to leave his father in these very delicate moments. But then he understood that it was more important to attend the meeting with his Maestra. She would know why she was asking that of him. Just when he decided to travel to the city, the doctor who was treating his father called to tell him that he had contacted a specialist in Paris and had arranged an appointment with him for Thursday afternoon. And so it was that Diego left Remoulins on the fast train that very morning reflecting on the singular way in which everything had played out.

Diego spent more than an hour speaking with the doctor. The specialist explained to him in detail that Esteban's condition wasn't going to improve if he didn't receive a bone marrow transplant. He stressed that given the haste of the situation, only Diego could be the donor. Now in the taxi, Diego contemplated all that the doctor explained to him and what he had learned during this time from Helena, trying to resolve the inner conflict that he felt in order to find an answer to his dilemma. He saw himself enveloped in a whirlwind of emotions, confusions and expectations that only collided in discord without finding a resolution. He made himself comfortable on the seat of the taxi, closed his eyes and wished for the Voluntad (Will) to be carried out.

At the other extreme of the garden of the Palais-Royal a tall, thin man leaned into the central patio and looked around. The glow of the small lights that bordered the garden filled the place with an air of joy. He wanted a moment of stillness in order to center himself before entering. He took out his pack of cigarettes and lit one, inhaling the smoke deeply to calm himself. In that instant, he saw Helena seated at a table. She was distracted arranging a scarf about her neck. He observed her, feeling a wave of emotion at seeing her. He was amazed at her appearance; physically she was almost unchanged. Her hair was still long and abundant, with a slight tone of mahogany. She moved gracefully, as always, but in her happy face he noted a strength and a presence that he hadn't seen before.

He felt exposed when Helena looked up and caught him watching her. She smiled and waved at him to come over. When Max reached the table, they greeted each other with an affectionate hug.

"Helena, what a pleasure to see you again." Max's voice expressed all of the emotion he felt at being with her once more. "But you haven't changed a bit; you look incredible. I can't believe that so many years have gone by." Standing in front of her he recognized the gaze that had always disquieted him. He remembered how sometimes that gaze had shown gentleness and others a penetrating intensity. Her green, almond-shaped eyes, open and frank, that contemplated everything with great vivaciousness, shone with happiness at seeing him.

"Max, what joy to find you again," exclaimed Helena with a smile. Max grinned and pulled over a chair to sit down. Trying to make himself comfortable, his long legs banged against the table and made it wobble. The waiter came over and he asked for a double espresso. Helena continued, "I was remembering our years in the University. I still laugh when I think of the pranks we played! Poor Doctor Garnier! I'll never forget the expression on his face that day in the lab! The two laughed at the memory. Helena added, "I recall how dedicated you were to your coursework. You had a determination to discover something important and you didn't let anything or anybody stop you. I wondered if you had invented the wheel."

"Yes, I've always wanted to find the key for curbing disease, but what can I tell you Helena, things haven't turned out like I hoped." He became somber when he said it, thinking of what he had experienced lately. Suddenly changing the subject so that he wouldn't lose control of his emotions, he said, "You know, many times I've remembered you and the things you talked to me about. I never met anyone that spoke and thought like you." Max lit another cigarette, but remained silent; he didn't know where to begin, he had so many ideas in his head. He saw the shopping bags on the chair and laughed, saying, "Some things never change, do they?"

Helena replied, playing along. "No, that's never going to change, Max. Do you like the color of my scarf?"

Max grinned and responded, "Yes, it pretty much blinded me. But tell me, we all wanted to know, what happened to you when you left Paris? Someone told me you were in Mexico. And what did you do there? How long ago did you return?"

"Yes, I was in Mexico for several years. Now I live close to Avignon." She smiled, saying, "The truth is that I've had many changes in my life and I've been in many places, but that's a long story. Tell me about yourself, Max. Judging by the bags under your eyes I see that you're going through something very intense."

Max played with the pack of cigarettes, turning it over and over in his hand while he listened to Helena. Looking into her eyes, he felt that everything that he had kept inside during the last few months was going to pour out as if he were a dam on the brink of overflowing. He began to recount his story feigning a distance and control which he didn't have.

Helena watched him and observed how difficult it was to sincerely express what's inside of us. It was like having a rock in a shoe that's bothering us, but we prefer to keep going without stopping to take it out.

"Well, I don't know if you remember, but I started to work in genetic oncology; then I got a position in the Research Institute here in Paris. You know, when I was young I had the illusion of being a great scientist who discovered the secrets of life and death. I believed that if I knew enough about genes, I could defeat cancer. But I don't understand why we can't win the battle against it. Some people get ill and others don't, even though they have the same genes. We have loose pieces of the puzzle which we aren't able to put together. Even though you won't believe it, I've even thought that there's something beyond us that decides or controls everything. I don't know if it's destiny or what." When he thought of Isabel, his anger and sadness spun round inside him wanting to come out and his voice began to break. He tried to calm himself watching the people stroll through the garden that was now illuminated by the lights. He confessed to Helena, "The worst part for me is that in spite of all of my resources, I couldn't even save my sister Isabel. You remember her? Almost a year ago she was diagnosed with leukemia and she died a little while ago, leaving her two young children alone. I did all I could to help her, but in the end I wasn't able to save her." He paused and sighed deeply. "The truth is that I'm furious because of everything that she suffered. I don't believe that she deserved to die that way and so young. I'm angry at myself for not having helped her." He remained silent for a few moments and ran his fingers through his hair.

Helena listened to him and noted his bitterness and frustration at having lost his supposed fight against death. She placed her hand gently on top of his to cheer him. She thought that Max's greatest frustration was because things hadn't turned out as he hoped, and not because of the death of his sister.

"And then, on top of all of that, several months ago I started dreaming about things that I don't understand; they're recurring nightmares that really unnerve me. I don't know why some fantasies bother me so much, but I can't find a moment's peace in my life." Max began to trace figures on the table with his finger, lost in his thoughts.

Helena observed the way in which Max played with his hands, trying to disperse the energy that was affecting him. She said to him softly, but with the intention that he draw out what he had inside, "I imagine that for Isabel, her illness and her suffering were something else, right? I suppose that she ended up accepting it all."

Without waiting for his answer she continued. "I know you feel very alone now and I understand the desire you had to help Isabel. But the anger that you feel at not having 'saved' her has to do with your reasons and not hers. Isabel's disease was part of her reasons, something that fell to her to live. If she was cured or not, it didn't depend on you or on anyone else. It's a fantasy, an act of arrogance to think that we have power over death; we can't avoid it, so stop reproaching yourself for it and we won't play at being god, which doesn't suit us at all. It's one of the great negativities of humanity; to try to avoid death and to live in the same matter forever."

"But that's my life, Helena. It's been my commitment to cure people so they can live as well as they possibly can."

"Look, science wants to make us 'immortal' with transplants, cybernetic organs and all of the other things that they're exploring, but we're unaware of the reasons that provoke our maladies. That's why we can't always cure them with drugs and surgeries. *Disease should not be fought as a surgeon does, who cuts and extracts, ignoring how the energy is acting in the man. When an organ or another part is extracted, there remains a blemish for the centuries in the Espíritu Huella. Illness is not of an organ but rather of the whole in general (18 Mar. 89).* The Espíritu Huella is like a 'register' in which all of our acts remain recorded and it's one of the nine layers that make up our matter. That's why almost all the illnesses we have in the present come from past incarnations. The body isn't like what they teach us in school."

Upon hearing these words Max became defensive and crossed his arms on his chest. He leaned back in his chair and said, a little annoyed, "What strange things you say, Helena. You who wanted to cure people and help them." He didn't want to think more about what her words implied, so he returned to his ideas. "Science only looks to improve the quality of life for everyone. With the advances that we're making, it's possible to live for more than a hundred years, and well. Are you going to tell me that this is something bad? Do you really want people to suffer?"

Helena watched him and thought, 'ah, how quickly he bared his teeth.' She calmly said to him, "It's not that I want people to suffer, Max, but that they understand. Death is simply a transformation that's necessary for everyone. Look, do you recall what I told you all those years ago about the Ego?" Max nodded his head in agreement. "Well, remember, the Ego doesn't die, it keeps on in its process of evolution and apprenticeship. All that we live, illness as much as death, is part of a process of development. But we should know that life is for understanding our state of being. *We are obligated to suffer so that we understand, and until we understand. [...]It should be understood that the Ego doesn't suffer deterioration, it is immutable, eternal, but it is induced in a programming, in an appearance, like a movie. [...]The frustration comes from not knowing who we are, or that this state of being is an appearance" (27 Mar. 89).*

Max relaxed a little, recalling the conversations that he had had with Helena. He remembered that it was he who sought her to ask for her advice. He decided to listen even though it might disturb him. He asked her with the curiosity that always won out, "Why do you say that we continue in an education in death, Helena? If we're already dead and we don't have a body, what is it that we're going to learn?"

"We live in a constant transformation, whether it be in matter or without it. The period in which the Ego is projected in matter is called Ansibir, and when it's detached from it, Trebolo. *When an Ego enters in to the state of Trebolo, it forgets what its material life has been, since it understands that this no longer has any meaning. Thus, it waits for a spark of lux (light) to which it should abandon itself and follow it in order to continue in its evolution. There are Egos that do not follow this lux as they are still fixated on that which was their Ansibir and therefore they do not advance on their path. They must*

wait for a second spark in order to advance although this is already weaker. This can repeat itself until the Ego loses its opportunity" (4 *Jul.* 87).

"And what is this lux that you speak of, Helena? Of course I've heard the tales of people who die and return, and say that they saw a light, or sometimes they pass through a tunnel, or see angels…"

"Ah, the process of Trebolo is very interesting. The Ego, when it separates itself from its matter in the moment of death, *sees a white lux (light) which is itself, it is the visualization of when it detaches itself from the matter, since it no longer has the care of this and all of its associated reasons (10 Oct. 87).* When it sees this lux, the Ego enters into it and feels a relief at not having matter.

To help it in its transition there are Entities that guide it to the place where it should go. The Ego perceives them according to its level of konsciousness and its desires or way of thinking. The Entities present themselves according to what the Ego wants to see in these moments so that it understands what's happening to it, since it lost its matter. How strange, right? Man sees what he wants to see. That's why some people say they see angels, others see their relatives and others still see things that are familiar to them. There are some Egos that still see themselves in their physical form and some others even reconstruct an environment equal to that which they knew because they don't want to let go of it. For this same reason it's important to let a person go when they're dying so that they search for that lux and they don't become confused at feeling the bond with the person or persons who do not want to let them go. *The Trebolo is an imagination. When the pequeño passes from one dimension to another he becomes saturated with new reasons. While the Ego doesn't change its understanding, it makes the world in the same way."* (19 Sep. 87).

Max, interested, asked her, "The Ego tre…, umm, the one who just died, does it have the same consciousness as in the Ansibir?"

Helena saw that Max's curiosity had made him forget his irritation and she answered, "Look, the Ego always has konsciousness of itself and can decide if it wants to understand or not. *The course of events of an Ansibir is like the projection of a movie. In the Trebolo, the Ego sees all of the projection of what its life was and many times 'skips' or blocks certain parts that it does not want to face and it will have to*

repeat the projection over and over until it understands (8 Apr. 89).
Therefore, if we have to live a situation many times, the fault is the
Ego's for being a lazy slacker."

They took advantage of the moment to laugh a little. They remained
watching the people walking in the garden; the afternoon was cool
and pleasant, and the place was full of activity. In the distance Helena
made out the figure of Diego, who came in rushed and nervous.

He walked up to the table almost running into it. She greeted
him with a smile, saying, "Ah, how good that you arrived. I want you
to meet an old classmate from the University. Diego, may I present
Max Gerard." Turning to look at Max, Helena introduced him to
Diego, noting his look of curiosity and surprise. Max extended
his hand, greeting Diego in a cordial but distant manner. A little
annoyed by the interruption he asked himself who was this boy who
had showed up to disrupt his chat with Helena. "Ay, Diego, what an
expression you have on your face. Look, what a coincidence. Max is
a specialist in questions of oncology and genetics. Maybe he can help
you. Would you like to tell him about your dad?" Helena wanted to
test Diego's reaction to see how secure he was in what he had been
studying.

They seated themselves at the table and Diego looked at Helena,
not understanding what was going on. Why was she introducing him
to this man? Could he help his dad? His emotions collided inside of
him again. He felt that he should leave things be, but the inertia of
his habits won out and he ended up asking Max about the possibilities
that existed for his father.

For his part, Max, although he didn't understand why Helena had
arranged this casual consultation, listened attentively to the details
that Diego supplied, proud of being able to demonstrate what he
knew. When Diego finished, Max remained silent a while analyzing
everything and he commented, "Well, what Doctor Manet says seems
correct to me, Diego. It's a terminal condition. But with this treatment
it's possible to extend his life, maybe three to five years. And it won't
affect you at all; it's a really simple procedure for you. Of course, your
father will need to be on strong medications for a while, and maybe
he won't recover his full strength, but he'll have more time to live."

Diego reflected on all that this implied. He sought Helena with his gaze, wanting to ask her what she thought.

Helena grasped his thoughts and said, "You have the means to find the answer for yourself. Ask what you should do."

Max watched them, not understanding to what they referred; he only perceived the force of Helena in that moment. Her green eyes shone with an interior light that he hadn't seen in the past. At that moment his attention was drawn to the emerald ring that she was wearing. It seemed familiar to him, although he didn't remember having seen it in the University.

While Max was distracted, Diego calmed himself and opened himself to ask by means of the Kommunication about what it was that he should do. Helena called the waiter over to ask for another coffee. Diego came out of his silence and turned a little in his chair to face Max.

"You know, Max, it's odd. I had wanted to find a solution for a long time, but now that I have it in my hands, I know that it's not the right thing to do. I understand that I shouldn't try to extend my father's life under these conditions. I believe that it's his time and besides, it's not up to me to decide this. I don't want to interfere in his process just because I feel sad or feel a commitment towards him or because I don't want him to suffer."

"But how can you make that decision young man? I don't understand you; didn't you want help? Don't you want your father to stay alive?"

"But it doesn't matter what I want, Max. The truth is that it's been very hard for me to understand and accept what I should do. I guess that you know about the obligations and responsibilities that a person feels for his family." Max felt a jolt of emotion within upon hearing this, recognizing what he had felt for Isabel. Diego continued, "I've looked for the way to cure him because I'm afraid of losing him. I understand that the relationship that we have with our loved ones goes beyond what we think." Diego, with all of the excitement that he felt at being able to talk about what he had been studying, exclaimed, "Look, Helena has taught me that there is no death, it's just a process of transformation. So if I want to change this, I harm him and myself.

I believe that there are things that we shouldn't mess around in. If an illness comes to us, it's because we need to learn something, right, Helena? And you know, I see him at peace with the idea of Trebolo. I think that it's been harder for me to see him and to let him go. Sure, he's not happy with me, because he says that he's going without me having settled down." Diego smiled, wanting to cheer himself up, but the truth was that he was sad.

Without being completely konscious of it, Diego had just taken a step towards breaking with a centuries-old programming that had been distancing him from his vital destiny. Throughout his lifetimes he had always had an excuse to put others before himself to avoid arriving at his destiny, involving himself with people.

Max looked within, recalling his own experience with Isabel. It seemed odd to him the calm and security that Diego demonstrated upon having accepted this resolution to the problem of his father. Max's bitterness showed itself in that moment. "Helena, I understand what you said about the Trebolo but now you're both talking to me about matter and organs and you lost me in these strange ideas." He remained quiet for a while trying to calm his thoughts. But inside of him there was an anger and a clash of ideas that didn't allow him to focus.

Diego took advantage of Max's silence to ask Helena, "The Ego in the Trebolo can see what it did in the Ansibir? An Ego can understand what it lived with clarity?"

Helena responded, "The Ego has intermittent moments of clarity in which it can perceive and understand what it lived and can see its faults and the things it did correctly. Its apprenticeship continues, but the same as in the Ansibir, everything is according to its level of konsciousness and evolution. *When it remembers the past it is experiencing it as intensely and completely as it lived it. It cannot be said in this moment it is dead because it is living it intensely. Of its past, an 'edition' is made, like in a movie, of what serves the Ego, so that it does not become saturated by what it no longer needs*" (3 Oct. 87).

Max began to come out of his irritation sufficiently to comment, "Isabel lived well; she was a good person, why did she have to become ill? Why do you believe she had to learn that way?"

Helena answered, "You know what, Max? In one way or another we are always receiving opportunities to understand ourselves, but we don't recognize them and we let them pass by because of apathy or rebelliousness. The Trebolo is a compilation of what was lived, but it's also a preparation for what is coming in our next Ansibir. In that state the Ego can decide, up to a certain point, how the conditions of its next stay in matter will be." She recounted a study to him. *"When the Ego is in the Trebolo it understands in an uncertain way; nevertheless, it chooses the circumstances that will surround it in the future Ansibir for its evolution. Sometimes it chooses to surround itself with many circumstantial acts and few that will diminish its karma, although some prefer to fill their lives with great hardships and sufferings, in order to purge their karma more rapidly. There are times in the Ansibir when it cannot deal with the burden and it is forgiven certain faults, the lesser ones and those that prevent it from fulfilling the greater ones. In the Trebolo a judgment is made of the acts of each Ego carried out in the Ansibir and only those that led it to its evolution are taken as favorable. That is why the Ego that chooses to be on the path of evolution is forgiven all those pending matters that it had in its Ansibir and these are not taken into account at the moment of judgment (31 Jan. 91).* Well, this last point is only if they are fulfilled.

I'll give you an example. A person who has had a previous spiritual advance, or who left something uncompleted because she was full of commitments in the Ansibir, can decide in the Trebolo that she isn't going to get married or isn't going to be rich so that she doesn't get caught up in reasons that distance her from her mission, and thus she can dedicate herself more fully to her own development. In the same way she can decide to have certain graces, like beauty, up to the point that corresponds to her. We make these decisions in the Trebolo but then in the Ansibir we no longer have the konsciousness of them, because we go through a process of forgetting before being born. That's why we don't understand why it is that we don't find satisfaction and we're always wanting that which we don't have."

Max considered what this implied. "If I understand you correctly, Helena, you're saying that it's possible that Isabel chose to become ill as part of a learning process. Do you really think that's the case? But what do you think she would have learned?"

"Well yes, Max, that may have been Isabel's case, or it also could have been due to a karmic reason. Maybe she damaged her matter in another life and this reason remained pending in her. These situations lead us to understanding our past which presents itself again in the now so that we can settle them. Normally the Ego is presented with situations that take it out of its everyday routine so that it questions itself. It's in that moment that we should act and modify what we need to correct. This is difficult to accept, I know, but it's part of opening ourselves to the truth of our existence. We are conditioned or programmed by so many previous life experiences that we allow ourselves to be carried along by our impulses; we aren't even konscious that there is a Voluntad (Will) and Kosmic Laws that govern us."

Diego felt Max's sadness, observing the difficulty that he had in accepting what Helena was telling him. But he wanted to understand more. Curious to be able to find a clue that would allow him to understand his own motives, he asked, "And why don't we remember what we ourselves program in the Trebolo?"

"This process of forgetting is part of the current laws of the Trebolo, but there is another reason which is due to the actions of men from other times. In the very distant past there existed a continuity between Ansibir, Trebolo and dreaming. What happened? Of the few who had the Konocimiento, some of them provoked an alteration in the matter that broke this continuity." She looked at Max and asked him, "What do you think, Max?"

Max didn't respond. He didn't understand why she asked him that. It seemed that she was trying to insinuate that he knew something about it, but how could that be possible? He felt as though his abdomen contracted and it became difficult to breathe. He was very uncomfortable and anxious. He inhaled slowly trying to calm himself. Inexplicably, he felt these words deep inside and for some reason that he didn't fathom, he felt guilty. He made an effort to understand where this feeling came from, but when he tried, it just vanished. He said to himself, 'but why is she telling me this if it doesn't have anything to do with me? I've always tried to do the right thing.' He leaned back in his chair and stuck his hands in the pockets of his coat, uneasy. He wanted to go; he felt a tension in his all of his body and in his neck. He didn't understand his reaction to Helena's words.

Diego was shocked by Max's changes of mood and by what Helena was saying. He asked himself, 'alteration? What could have happened to change the sequence between Ansibir, dreaming and Trebolo so drastically?'

Max said to them, more abruptly than he wanted, "Pardon me, Helena, but I'm a little tired and you left me with a lot to think about. Would it be possible to continue talking another day?"

Helena observed Max's agitation. She knew that the dreams that weighed so heavily upon him were a key that would permit him to understand something vital about himself. They were linked to his past, to the death of his sister and to the anger he felt at not being able to cure her. She turned to look at him and said, "Yes, fine, Max. I have a little free time in the morning. Is nine o'clock at the octagonal fountain in the Tuileries Gardens ok with you? I've always loved that spot. And you need to wipe that look off your face, you're going to frighten the people on the street," she said to him with a smile.

Max bore up and said goodbye to her with a kiss on the cheek. He held out his hand to Diego saying something about it having been a pleasure.

Diego watched him leave and said to Helena, "Would you like to go walking for a while? The night's very pleasant."

They strolled along the streets in silence, enjoying the evening. After a few minutes Diego commented, "Wow, Max has such a deep despair. What torments him so much?"

"His past. It's the moment when it should come out, but he won't allow it. You know, being at this point is very delicate; there are two very great forces that consume a person. There's the impulse that wants the memory to come out and the other that wants to hide it. But I believe he won't be able to keep covering it up any longer…"

Diego remembered something Helena had told him earlier that he wanted to clarify. He asked,

"Why did you say that we don't remember what we understand or program in the Trebolo? You pointed out something about mutations and something that changed the sequence."

"Yes but it's very complicated and I'll tell you about it at another time, Diego. Or maybe Max will. Well, for now I can tell you that each Ego *forgets everything a month before leaving the womb; it enters into* THE REASONS OF OLVIDO (FORGETFULNESS) *and remembers who it is again at 7, 8 or 9 years of age, according to its evolution and reasons (21 Jun. 90).* This is so that the reasons of its past don't provoke an imbalance in it. We don't currently have the capacity to understand and incorporate all of this information without it causing disorders in us. Nevertheless, it won't always be like this. It is a truth that we will be able to remember our previous lives clearly. In the future we are going to have the konsciousness of all that we have lived so that we can resolve what is still pending. If we study, we are going to have this sequence between Ansibir and Trebolo. It's part of the changes programmed for humanity. Can you imagine it? You're going to be able to enter into matter knowing what you have to do and why, if you develop your konsciousness and if you advance in this life." Diego was astonished, imagining the great possibility of being able to retain all of this information. 159

Helena spoke again. "I want to say something to you, Diego. The tests for each person are very different. In your case, this first step is to clean up your relationship with your father, but there are many more to come. We have to leave behind everything that impedes our evolution." She thought of Max and of the tests that he was facing at this moment. Could he vanquish his anger in order to take the necessary steps? She took out a cigarette and lit it, asking Diego, "And now, tell me. When are you thinking of returning to Remoulins?"

Chapter 9
Memories

*Another cripple declared, "There are no superior beings,
everything is chance and everything is happenstance."
What is chance and where does the flower
come from if there is no seed?*

RONRONEO (PURR)

Although the drizzle continued falling softly over Paris, the fleeting rays of the sun managed to emerge from between the clouds making the city shine. Helena entered the Tuileries Gardens and walked towards the enormous octagonal fountain that dominated the space. The trees moved with the breeze and she felt the wintery cold typical of February in the city.

When she arrived at the fountain she saw that Max had not yet appeared. The rain stopped and she sat on a bench to wait for him. The sound of the water falling in the fountain disguised the sound of the morning traffic. She thought about Max and asked herself how he would have passed the night. What she told him was only the beginning of the things that he would have to see if he really wanted to find his way out of the hole he was in. She said to herself, 'Max has lost his openness; now he's very skeptical, he has a very structured mind and he wants to prove everything with evidence and tests. Before, he was more intuitive and more amenable to accepting new ideas. I hope that he can open himself up to listen, since he's accustomed to imposing his speculations. But for some reason this opportunity is being presented to him...'

Five minutes later Max arrived. In his hands he carried two cups of coffee and a bag of hot croissants. When she saw the circles under his eyes Helena knew that he hadn't slept well. Nevertheless, Max greeted her with a smile.

"Good morning, Helena. How do you like the cold? I brought something to warm us up a bit. I remember how you liked coffee."

Helena stood and took the coffee, feeling the warmth between her hands. In spite of his apparent kindness, she noted a distance and anxiety in him. The two walked slowly through the Grand Allée towards to Louvre Museum, eating the croissants and chatting.

"I've been thinking a lot about what you told me yesterday, Helena. Look, I feel more at ease about Isabel and her death, but I'm not in agreement with everything you said. It's just that you talk to me about things that can't be proven." He stopped, deliberating. Helena waited silently, attentive to what he was going to say. Knowing him, she knew that his first reaction would be to argue about what she had told him yesterday. Max continued, "I believe that medicine has been a great aid in order to be able to avoid suffering. However, you say that sometimes it's necessary to suffer. Well, if I accept that part I'd have to ask you how it's possible to know if it's necessary or not. Yesterday it shocked me how that boy Diego arrived at his decision, even though it's obvious that it still affects him. He seems very young to have that kind of confidence."

"Young? There's nobody young here, we're all way past our prime." Max didn't know what to make of her remark, but Helena laughed and continued. "For you, what is suffering?"

"Well, it's obvious; a pain, torment or stress that…"

Helena stopped him and said, "Within the reasons that we know, suffering is a pain that we want to avoid at all costs, right? But in Kosmic terms, *suffering is a field of energy (positive or negative to our understanding) for a transformation and an understanding. We are not given what we want, but what we need. The submission to the Kosmic reasons facilitates these necessary changes (14 May 88).* There is a Kosmic order which is constantly dictating changes to us so that there is growth. Suffering is necessary so that we find an adjustment within ourselves and a place within the Kosmic. When we don't resist the changes, the pain and suffering are less. Everything within Life participates in an arrangement of laws that govern it; not knowing them doesn't mean we can break or ignore them. In short, we should ask permission and accept what is indicated to us without worrying if a suffering or an improvement comes as a consequence of it. But we always have an interest and we want to avoid the change, and this is what provokes the suffering.

Look, within the reasons of the Universe or Fusion, we humans are incipient particles, we're barely learning to 'walk,' and as such, we need to be guided and taught. We're like little children that need to walk holding onto their father's hand. But we don't accept this relationship and we always want to impose what we want, without asking ourselves if it's correct or not. This has made our isolation here in the Planicio Terráqueo greater and that's why we feel estranged from the multiple reasons that surround us and guide us. Nevertheless, we need to develop this relationship again. Thus as my Maestra told me once, *'[...]A man wants to know the time. If he does not ask he cannot know it, but if he approaches someone and asks them with humility and respect, "what time is it?" he will be answered'" (Nov. 10).*

"Why do you say that we need to ask permission? Permission of whom or for what? You're not going to tell me that god is watching us."

"As a first step, we should ask permission of the Perespíritu, which is a guide that every Ego has. I'll tell you more about this later. Well, in regard to your comment, I'll tell you that the image we have of a 'god' that created us isn't correct. Yes, there's an order in everything and there's governing Entity of the Kosmo that we inhabit, that we know as the Omnipotente Padre Eterno Jehovah (Omnipotent Eternal Father Jehovah). *The Planicio Terráqueo was ordered and conceived by our Padre Eterno to whom we owe respect and submission" (11 Apr. 11).* Helena saw the look of disbelief on Max's face. "And get rid of that image in your mind of the old man with a beard, it's not like that. He is a Koncepto that planned the development that we'll have on the Earth. He didn't create us; he formed the Planicio Terráqueo and the matter that we have here. You see, the Ego wasn't created, but was manifested and then formed by the Supreme Entities. Why is it that when we think of a Supreme Entity or Intelligence that we want to imagine it in our own image?" she asked. "We always want to lower the greatest to our level instead of elevating ourselves to know what is beyond our awareness. With much effort and sincerity, a pequeño, or little one, can develop his relationship again with the Omnipotente Padre Eterno."

Max didn't know how to respond to these words so he remained silent. Skeptical, he put Helena's words to one side; he didn't want to think of their implications at this point. However, his scientific mind wanted to know more about the development of the planet, but he was afraid to open up the subject with Helena. He thought, 'who knows what else she'll come up with now!' They walked a bit more watching the activity that filled the park: children running and playing, other people rushed to get to work, and a pair of teenagers that wandered peacefully, absorbed in their own world.

"Yes, it truly is a difficult subject, but it's better if you tell me what's disturbing you so much, Max," Helena said, grasping what he was thinking.

Max ran his hands through his hair, trying to focus his mind. "Ok, Helena, there's one thing that you mentioned to me yesterday that really unsettled me, even though it seems like nonsense to me. You said something about mutations that were carried out in the past. I don't know why, but it affected me deeply. Do you remember that I mentioned something to you about some nightmares that haven't left me in peace? Well last night I dreamed of them again. Maybe you're going to think that I'm an idiot, and that they're just my imagination…" He was ashamed to think that these illogical dreams were interfering with his ability to function.

"Don't disregard your dreams, they're not fantasies. They come in precise moments; they're lessons. Of course, sometimes they present themselves in a confusing way, or surrounded by a different reality than what we perceive. The fear that you feel is because they're touching something that's stored away inside of you. Why don't you tell me about them?" Helena stopped and sat on one of the benches in the park.

Max sat down as well and took his cigarettes out of his coat pocket. He lit one and inhaled the smoke to extend the time. He was unsure of how Helena would understand what he was going to tell her, but he plunged in and began to recount his dreams to her, as if he were explaining one of his research projects.

"They began about four months ago and they recur all the time. Maybe you're going to think I'm half crazy because they're such strange things. In one dream I see myself, not as I am now, but even so, I know that it's me. My body is different, it's lighter, translucent. I'm in a place full of light and brilliant colors. I realize that I'm perceiving the environment with all of my body: the sounds, the colors, the energies of the Earth, the smells. I feel everything that's happening around me, I'm in contact with the rest of the people, with the animals and the plants; I don't quite know how to explain it. We understand ourselves with pure thought. And then the scene changes. I see images of a terrible destruction, everything has disappeared and the world has been left in darkness, there is no light, and the people who have survived feel a great fear. They're terrified, traumatized, because earlier everything was full of light and now there's only darkness, everything was filled with shadows. They don't know what to do. They can no longer think clearly or remember who they are. I realize that my body is no longer in contact with the environment or with the other people around me and I feel lost. I have the sensation that something was separated from me and it hurts me psychically. I know that now we are in a more solid world, harder, and that I've lost an important part of me but I don't know what it is. I wake up sweating and with a terrible anguish that I can't describe to you."

164

Max remained quiet, uneasy because of the sensations that the dream still caused in him. "Well, when I started dreaming these scenes I put myself to look into dreams in general." Max explained his long and boring investigation. To conclude, he said to Helena, "Frankly, of what really happens during the process, not much is explained and almost no one gives it importance, deducing that they're fantasies. But in some dreams I see myself in another time, Helena. If they are, as they say, just fantasies, then where am I getting this information from? And if they're not fantasies, how is it that I see myself in another era? I don't think that I'm imagining it. And the things I see, I don't know how to explain them to you. Sometimes they're incredibly beautiful and sometimes they're terrifying scenes. What do you think about all of this, Helena?"

It seemed that the clinical analysis that he had just presented didn't take any of the weight off his mind. The disquiet and worry that his dreams produced in him were obvious.

When Max had finished with his chatter, Helena replied, "There are many things that aren't understood through science. Why? Not everything can be reduced to physical and chemical processes. Man sees matter, perceives matter, and then takes it as fact that all is matter. We don't give importance to dreams because they occur outside of the physical. There are many reasons we don't discern because our konsciousness has changed. Do you remember when I was dreaming of many strange scenes when we were in school?" Max assented, moving his head slowly, waiting for what Helena was going to say. "Well I also saw myself in another time, in places and with people that I didn't know. And the things that I saw disturbed me very much. But when I began to study, I was able to understand what they were about and how it was possible to have those memories. Of course, in that moment they weren't precise, but they had a truth. What you see also has veracity, Max.

Look, a dream isn't a fantasy that we re-create, but a perception that's different from that of wakefulness. ***What we understand as dreams are in reality levels of konsciousness, and according to the comprehension of the Ego, the level in which finds itself. When a pequeño 'sleeps' he projects himself in different levels of konsciousness"*** *[…] (18 Nov. 91).*

"I don't understand what these dreams have to do with my past but yes, I live them as if they were a reality."

"We tend to think that wakefulness is our 'reality' and that dreaming is an imagination. But the truth is that dreaming is a parallel projection of the Ego, in which it continues in an apprenticeship. There we are as 'alive' and as konscious as we are in this moment. In fact, it's more real than wakefulness."

Max looked at her, incredulous at her words. "You're not serious are you, Helena? How can it be more real than this?" he said, touching the bench with his hand. "When we're awake we have control over what we're doing, we have continuity during the day, time is linear. On the other hand, dreams are like events or images that arrive randomly, there is no sequence, we don't have control; they aren't real." When he said this, he realized that he was contradicting himself.

"And if your dreams are only fantasies, then they can't hurt you, right? Nevertheless, I see that they're really disturbing you." Helena took a sip of coffee and asked him, "Do you really want to understand something about your dreams, Max? Or are you going to insist on seeing the world in the same way as always? Ok, I'll tell you a part of what I've understood about dreams through many years of study. It's something that the first men understood and utilized, but this Konocimiento has been forgotten in the centuries that have passed. It's because of the changes in konsciousness that I told you about yesterday." Max made an effort to open himself up to hearing what Helena wanted to explain to him.

"Dreaming has many facets and is very complex because it is developed in a level of awareness that is different from that of wakefulness. As scientists say, there are physical reasons for which we dream, but they don't know that there are other, deeper reasons that have to do with the conformation and evolution of the Ego. When we sleep, the body rests and the Ego detaches itself a little from the demands of the matter and can project itself to different levels of konsciousness. During the day we fill ourselves with residues, as much emotional and mental as those that stem from the wearing away of the matter itself. Dreaming helps us clean ourselves of this. So then, it is a rest for the matter and a nourishment for the Ego when it achieves a projection of itself to more complete levels. Within the Astral there are different levels or strata. The lowest are impregnated with the reasons of daily living, dense thoughts, chaotic emotions. On the other hand, in the highest strata the Ego is more detached from its habitual environment and can grasp other reasons that are not dictated by the interests of matter. That's why we sometimes perceive something that is going to happen or something that already happened. *Dreaming is the saturation of thoughts and actions that are established by memories, immediate actions and yearnings; this implies a saturation and the Dream is a discharge of what is not considered important of the images transmitted to the Brain; it is therefore a form of exhaustion in which the neurons are harmonized and for example the organism discards the intestinal residues (5 Sep. 10).*

Then, if you really think about it, you'll see that just the same as in the Trebolo, dreaming is another state of konsciousness. For example, **when a person sleeps, he passes to other dimensions, not always in plenitude. Life doesn't end when we die, it is a dimensional change"** **(3 Oct. 87).**

Max interrupted her. "Ok, I understand what you're saying, that wakefulness and dreaming are different states of konsciousness." He stopped, perceiving something more profound. "Wow, wait a minute…if I understand you correctly, then…"

"Wakefulness, dreaming, and Trebolo are a continuum in the projection of the Ego. Yes, they are different states of konsciousness in which we don't always experience it in plenitude, but the Ego is always konscious."

"Now I understand why you say that we live in a continuous apprenticeship, but that we're not aware of it." Max was impressed to see it like that. Reflecting on it a little more, he perked up, grasping the possibilities of what this implied. "Would it be possible to arrive at that continuity in konsciousness? If we had it before, can we have it again?"

Helena listened to him, observing his duality. Max had an eagerness to want to know more and he grasped what Helena told him fairly well, even though she knew that he paid attention only to what suited him. His main interest was to have information in order to control or modify what affected him.

"Yes, Max, in fact we should arrive at the point where we live with a continuous konsciousness within the states of wakefulness, dreaming, and Trebolo so that we have continuity in our education. With this continuity we'll be capable of remembering what we agreed to fulfill upon entering into matter and of what we have lived in the past. Thus we will have the certainty of what falls to us to live in each incarnation. It'll be easier to complete this, since everything will be a marvelous sequence that will lead us to our final destiny: completing our lesson on the Earth and incorporating ourselves again into the magnificent development of the Universe or Fusion. The continuity that we should search for is in our konsciousness, in developing the mind. This doesn't have anything to do with intelligence, the desire to dominate the Universe, or with practices that seek to control the projection in the Astral in order to take a power from it. Rather, it's the acceptance and understanding of what we are and that we belong to the Fusion, as if we were part of a huge family."

For an instant it seemed as though everything in the park became calm. A few large drops of water fell on the bench announcing the arrival of a downpour. Max's question remained suspended in the air while the rain began to fall with more force. Helena stood up and said, "Let's go! We're just a few steps from the Louvre and we can keep talking there."

They ran to escape from the torrent and entered the museum. They left their wet coats in the cloakroom and began to view the exhibits. They enjoyed a few moments of indulgence while looking at the artworks. Upon arriving at the sculpture salon they made themselves comfortable on a bench. Max, anxious to understand the scenes that he had seen in his dreams, asked, "Then how do you explain those strange images that I'm seeing?"

Helena was surprised that Max had dreamed of those very remote images, managing to see part of the experiences of the First HTime and one of the destructions that occurred in those faraway eras. They were recollections that were buried deep in the man's memory, difficult to perceive. They were emerging into the light in these moments so that he could face them. She sensed a great desperation in him and understood his pain. Looking into his blue eyes which in that instant expressed an ancient sadness, Helena touched his arm and explained, "What you're seeing are reasons of humanity's past. They are very, very old memories. After those events it was necessary to erase certain experiences from the mind of man so that he wouldn't cause more damage to himself. And now man only has the remembrance of one day having been expelled from a paradise. And why does it fall to you to see them? Because they're reasons that are part of your own actions, something that you lived and you remained marked by them.

Our history isn't simple, but with a little or a lot of effort we'll be able to understand it bit by bit. When these memories surface and we comprehend them, there comes a great liberation. But sometimes they make us feel ashamed and we don't want to see them. It's better to release them and not to carry them; they're very heavy, aren't they?"

Intrigued, Max asked her, "Then what I saw really happened?"

"Yes. *In the beginning the Earth was a paradise in which man could freely wander about, he fed himself from an atmosphere not yet contaminated by his negative thought. He utilized his faculties to not suffer the environment, regulating the temperature of his body by activating his internal energy nuclei. [...]He could communicate with the animals. He was in relationship with levels superior to him and he lived in a deep contemplation for his apprenticeship. He had no need to travel as his faculties allowed him to teletransport himself. But he rejected that world because he wanted to form his own and he invented a fantasy, a modification of libre albedrio (free will) that has led him to the destruction of the environment in which he lives, and to a deformation of what was the ideal man. Now he alters that which he is unaware of. For example: with 'medicine' he altered the balance: he made an effort to reduce the death rate and now he struggles to control overpopulation, and suffers the distortions that he caused in his matter, it was not made to contain excrement, or so that a woman menstruate.*

The structure of man was exact, he was in harmony because of the submission to that which created him, and the modified libre albedrio distanced him from this principle" (9 Jul. 88). 169

Max looked at Helena fixedly and said, "How sad, Helena. You talk of a past full of beauty and harmony; how did we lose it?"

"Look, your dream speaks of two reasons: of that 'paradise' in which we lived and of what happened when the balance was broken and mankind remained in the darkness. When the reasons of mankind and of the Planicio Terráqueo were distorted by man himself, he was warned but he paid no attention. So then there came cataclysms so that he wouldn't cause more damage than what he already had.

There have been multiple destructions over the course of our history and each one has represented a step backward for humanity. The Konceptos limited us in our graces. In this way we lost the memory of everything and we started over again, but with fewer abilities. Our actions have brought as consequences alterations in our matter, limitations in our konsciousness and in the way we project ourselves; we lost the continuity between wakefulness, dreaming, and

Trebolo. Not to mention the destruction of the environment and the diseases that have developed. Worst of all is that now the reasons of matter dominate the reasons of the Ego. The men of the first eras were more apt, their konsciousness was more developed and they lived without the illnesses and the material worries that weigh upon us now."

Max remained silent for a long while; what he was hearing was too much for him. He was making a huge effort to control his feelings. He had dared to recount his dream and it had left him in pieces. It was a small instant in which he stopped posturing. Helena perceived in Max an Ego that has always carried the burden of great pain and suffering at knowing himself to be part of these actions; even though he wasn't konscious of it. She wished that this moment of openness would grant Max the opportunity to overcome his fears and that it would give him the strength to advance upon his path. But she knew that it would pass quickly and Max would once again put on the mask he had worn for centuries.

In these moments Max was really uncomfortable with himself. In the past, the actions of a few had provoked changes in the matter and in the konsciousness of man. Max had been part of these past actions and because of them he felt a great sense of guilt, even though he didn't understand why. Part of him wanted to argue, but the other part wanted to silence itself, at feeling the weight of Helena's words. He said, "In the dream I feel alone, abandoned, with a sorrow and a feeling of guilt so deep that I can't explain it to you. What do I do? How can I stop these dreams?" In this instant Max lowered his guard and spoke sincerely to Helena.

Helena touched his shoulder, saying, "Easy, Max. The important thing is not to stop them; it's to understand them. These memories are an opportunity for you to clarify your reasons. If you want to understand, Max, ask for it. Give yourself this chance."

Max began to pace around the exhibition hall, trying to distance himself from the sensations that smothered him. When he calmed down a little, he sat beside Helena. Evading what he had been feeling, he changed the subject, entering once more into his role as scientist, and asked her, "Is there a mechanism that allows one to see the past in dreams?"

Helena stretched her legs and, observing Max's facial expression, rubbed the mole on her forehead.

"Ah, well that's a very interesting question, Max. It turns out that everything that we've lived throughout our incarnations or Ansibirs, and even before arriving to the Earth, is stored within us." Upon seeing Max's surprise, she affirmed, "Yes, we have a 'record' inside of us in which is registered every thought, every word and each one of our actions. It's not in the brain or in the blood, as some believe, but in a layer of the matter that's called the Espíritu Huella. *The Espíritu Huella represents in itself all of the manifestations or actions since the beginning in which the Ego entered into the Reason of Life. […]* It is that *which throughout time was called 'Astral Body' (7 Nov. 08).* And that's how part of this record comes to us in our dreams, sometimes in a gentle way and in others in a chaotic way. In the moments when we enter into a deep dream, we can enter into contact with this information according to our level of konsciousness and the interest that we have. This record can help us to find our past reasons in order to examine and understand them.

Of the past it forms certain habits that are registered in other incarnations. It is thus how the brain *combines, without realizing, the past reasons that surface tenuously and which are projected by the Espíritu Huella. This is what thou callest ensoñación** (*the state between dreaming and wakefulness). *In dreaming, the matter is transformed by means of the brain, as much as it is able to, in adjustments. The Ego delivers the connection that tries to be in accord, and the Perespíritu delivers imperative reasons in the reaffirmation of the vital or successful structure of the Ego in matter. The Perespíritu stimulates the Ego, and motivates the matter to the order and respect that it owes the Ego (13 Jan. 94).* That is, in the dream, there is a great adjustment between the reasons of the Ego, those of the matter and those of the Espíritu Huella so that the reasons of the past emerge and can be understood. This adjustment also allows that the Imperante* reasons (*prevailing rules) of the present become visible to the Ego. All of this process occurs through the brain, which is like a translator. The Perespíritu helps coordinate the reasons between the Espíritu Huella, the Matter, and the Ego so that they arrive at plenitude.

The Perespíritu has a very important role. *It is a controller of our actions and of our feeling and perception; it can and should relate itself in an effective way in us in order to understand the Espíritu Huella (20 Jan. 09).* It helps us connect ourselves with the reasons that govern our development. In our dreams, if we ask its help, it can indicate to us what is vital and what we should do, and in that way adjust the reasons that are contrary to or favorable for our evolution. This is one of the lessons that the Maestro gave us, but we need the permission of that which created us to be able to achieve it."

Max, enthused, put aside momentarily that which didn't fit within his scientific explanations regarding the body and exclaimed, "Wow, then there really is a lot that we don't know about the process of dreaming. But it seems incredible to me to think that we can arrive at an understanding of our past that way. I never would have thought it possible that all of that information was inside of us, but yes, it would explain a lot..." Returning to reflect on the images that he had seen, he questioned her, "And of the things that were done in the past to affect the matter and the konsciousness, what can you tell me?" Max said, running his hands nervously through his hair, as was his custom.

"Many centuries ago, when the world and man were different, we received teachings about matter, the environment, and our own nature, among other lessons, so that we could understand why we were here in the Planicio Terráqueo, where we come from, and to where we need to arrive.

In the distant past man was given an opportunity to become aware of the Entities that regulate the Kosmo which are responsible for promoting the activities in it and in the Planicio Terráqueo, like the winds, storms, earthquakes, the solar and lunar energies, and which also regulate in man his organs, neuronal activities, karma, and more. As a reference they were given the name of Genios. *They indeed have a worthy affinity with and understanding of the human genus, and also more often than not they approach those pequeños who have the determination to know, in order to deliver their grace. They*

themselves are the mysterious occult masters (14 Jun. 96). Their teachings form what man has called 'Magic.'"

"Helena, please, magic is something for children."

"And for witches, too..." Helena laughed. "Well, then, you're going to like this, Max. *Magic is a way to be able to perceive that which is not structured in the mind of man in that instant (26 Mar. 88). This greatness was delivered to the men of the first ages, since within them prevailed nobleness and serenity. But these men, upon losing their purity, lost this faculty or grace, and thus, this reason was transformed, and changed into science, but these faculties were not used to promote grace, but rather disenchantment and the destruction of man himself" (14 Jun. 96).*

"And if these Entities are on the Earth, how is it that we don't know about them?"

"We don't sense them as we did before because this possibility was taken away from us when we used their teachings wrongly. Their projection is in a level that we no longer perceive. Nevertheless, they are still fulfilling their functions here and in the Kosmo."

All of this was breaking with Max's mental structure and he battled not to reject it. He abruptly exclaimed, "Then you're saying that science comes from magic? Isn't that going a little far?"

"Magic, that is, the teachings of the Genios, turned into science when man began to use it to control and to modify, in order to impose his interests of dominion, and not for the education and recreation for which it was given to him. Men wanted to imitate and reproduce the wonders of the Genios. They wanted to create, manipulate and control matter, the mind, and the environment. For example, they tried to transform the gene in order to take the powers of the animals and incorporate them into man. *Man thought to obtain power over life and altered the genetics making absurd combinations, monstrous mutations. His structure did not allow deterioration, oxidation. Now his matter breaks down continually. The true man was annihilated (16 Jul. 88).* Well, when the Wisdom, that is, the teachings of that time weren't used to arrive at a comprehension, they were restricted. Our perception was limited so that we wouldn't continue distorting ourselves. Wisdom remained as a memory in the mind of man, and thus continued to be used, transforming itself into what we now know as science. One of the consequences of these actions of the past is that the konsciousness of man is misadjusted, since it no longer clearly perceives his own reasons or the Konceptos. And there were other very grave consequences that still echo in the present affecting the current humanity.

Mankind in the present is repeating these mistakes with cloning, genetic manipulation, the search for antimatter and more. Of course the real intentions for this search are hidden; sometimes not even the scientists themselves know what they are. They are pawns in the game of the hierarchs."

"What?" exclaimed Max. "Aren't you exaggerating, Helena?"

"I wish. In the first eras, we lived in a continuous roaming between the Astral and the matter, between ensoñación and wakefulness. *Man imposed work upon himself when he was projected originally to contemplate and so, find his way. He contemplated and delighted himself, he ate little, he didn't act against the established order as he does now with medicine, he didn't break its own ecological reasons. The future humanity should return to the beginning, to the contemplative, but it already destroyed the environment. Man is a parody of what he should have been (2 Jul. 88).*

Man, upon wanting to modify his state of being, remained trapped in what he himself re-created. He inverted his interests and threw himself into the material world, wanting to take over the surroundings. All of this was because a few men wanted to dominate all of humanity. To achieve it, konsciousness was fixed in the material and it was thus that the necessities of the matter subjugated the necessities of the Ego, until it made our perception of the Ego almost inexistent *[...] since the material is simply an evolution that fills the senses, like a drug (17 Mar. 11).* That's why it's so hard for us to try to once again perceive what is not tangible. What I study are doors that allow us to perceive these reasons again and relate ourselves to the environment as was planned in the beginning.

With this change in our konsciousness, we now see science as a life saver, the maximum expression of man, that supposedly seeks to do good by studying the world, the body, the Universe, but the truth is something else. I'll give you an example: they try to help people when they can't have children. But not everyone should have them because there is a balance and Voluntad (Will) in this. Having a child is for a specific lesson that is not an imperative in all of our Ansibirs o incarnations. When it is time for an Ego to enter into matter, there is a process of accommodation that is dictated by Kosmic Law. There is an Entity which designates the matter that corresponds to each Ego according to what it needs to live. The place where it should be born is also decided. When this process is interfered with, the Egos that enter by imposition disrupt their own development and that of the rest. Why don't we accept the rules of our condition here and instead insist on doing what we want and not what we should? How many billions are we now? We're sabotaging ourselves."

Max couldn't take any more and exclaimed, "I can't believe that *you* are telling me this. You who wanted to help people and save the world. Science has propelled humanity to extents never before achieved. Science has led man to the utmost, we're better off than ever before. That's why I dedicated my entire life to finding improvements."

"Don't fool yourself, Max, what you want is recognition. If you really had an interest in helping, as you say, you'd have to think of the consequences. If you want to advance, you have to open yourself to new reasons that are beyond your understanding. Now you have to accept that not everything is within your possibilities and that you need permission and a guide in order to understand more. But if you want to approach these new reasons you need humility and submission."

Max's pride evaporated upon hearing these last words. He thought to himself, 'but I haven't sought greatness in my life, I've just wanted to help people. And submission to what and for what?'

Helena realized that Max's anger wouldn't allow him to accept more at that point and she said, "Come on; let's go see if it stopped raining. Look at that painting by Velázquez. What a painter! Isn't it fabulous?" They left the Louvre and walked in the direction of the subway station. Helena mentioned that she had to leave to finish her errands.

"Well, Helena, I'm going, too, I have a lot of things to do. I need to get back to my lab, to waste more time in my useless investigations."

"Yes, maybe you'll find a cure for your nightmares." Helena commented in a more serious tone, "You know what, Max? Intelligence can't get you out of this situation. You think that the mind is the maximum. But the mind doesn't transcend, it's a part of the projection of the Ego in matter. Everything that you know is Apparent, it's not Real. You can manipulate the Apparent, study it, pick it apart, but everything that man disrupts stays here in the Planicio Terráqueo, in the Apparent; he can't touch the true reasons of the Ego. All of your efforts won't lead you to know your true reasons because they come from the Real. In order to understand your truths you need the Konocimiento. Without It, you'll live a frustration that will only serve to confuse you more." She saw the anger in Max's face; his blue eyes darkened. In spite of this, she continued. "I'm going to tell you something else. You don't remember it, but once you made a commitment to use the teachings of the First HTime well. What did you do with them? And what are you doing now? This is what you should ask yourself. I'm curious about one thing, Max. Why did you call *me* to help you?"

Max remained mute; he couldn't say anything in the face of Helena's strong words. She said to him, "Well, I'm going, I wish you the best." Helena held out her hand and gave him an affectionate hug. She turned and walked down the stairs to enter the subway.

Helena imagined Max standing at the edge of a precipice. She knew that if he rejected this opportunity, it would mark a step backward in his evolution. It was important that an Ego like Max, so involved with the reasons of humanity, would comprehend because when one pequeño understands, he helps the rest to find the path.

Max watched her as she descended the stairs. The sounds pronounced by Helena provoked a great distress in him, and to calm himself he began to walk rapidly towards the Institute. He couldn't stop asking himself why she had indicated that he had a commitment pending. That idea, for some unknown reason, produced even more anxiety in him. Could it be that she was right? He couldn't get the words, "Why did you call me?" out of his mind. He decided to ignore his emotions and continued on towards the Institute. He said to himself, 'this has been a very unpleasant day. She's outrageous! The things Helena can come up with. She left me more upset than I was. But how odd, at the same time I feel attracted to her and to all that she says. Well, tomorrow I'll feel better.'

Chapter 10
Encounters

Man, upon avoiding the Konocimiento,
formed a parody of it, and this he called Wisdom,
giving himself over to it and with it,
avoiding the encounter with the Konocimiento,
leading him inevitably to a mistake.

KONOCIMIENTO

In the three months since his trip to Paris, Diego experienced many changes in his life. His father had died and, even though Diego understood that it was only a process of transformation, he sometimes felt the emptiness that his absence left. In order to try to rid himself of the confusion that his emotions provoked in him, he immersed himself in study with the motive of understanding the reasons that united them and the importance that there had been to separate himself from his father before he died. As Helena had underlined, "When you are initiated, the Konocimiento is delivered to you, which is the keys to be able to understand life. You will have to make the effort to understand how, throughout your incarnations, you have arrived at this point, and from here on in, how you should direct your actions in order to be able to synchronize yourself with the motives of your own evolution." She had added, "You need to be presented before the Konceptos (Koncepts) by a Maestro, who will initiate you and act as your guarantor. I can induct you on this road, but it will be through your surrender that you will find the guidance of the Konceptos, and if it is their Voluntad (Will), they will indicate to you where you should walk. All of this will be carried out by means of the question, the answer and analysis, and principally with your humility, submission, and respect to accept what they designate for you, that you will fulfill your destiny."

Through study, many of the circumstances of his life that previously had seemed unfair to him and filled him with resentment were clarified. Diego understood that each act, deed or reason had led him constantly towards the encounter with his destiny. Upon understanding and accepting what he had to live, it allowed for the arrival of new reasons in a harmonious manner.

It was May and the climate had changed; it was very pleasant. Diego took advantage of this to explore the countryside while he walked towards Helena's house. He enjoyed his outing, stopping in various spots to pick some poppies, daisies, and other wildflowers. The cheerful colors attracted his attention, and thinking of Helena, he decided to make a bouquet with them. When he arrived at the house, he knocked on the door and she called out from inside, telling him to let himself in. Diego saw that the darkroom light was on; she was finishing a job for a magazine. Helena walked out and greeted him, accepting the flowers with delight. She invited him to enter into the living room while she arranged the bouquet in a glass vase.

Diego prepared himself to study, removing his notebook from his backpack. He deliberated about how to phrase for Helena the question that had him uneasy all day. When she settled herself in her chair, Diego got the nerve to ask. In Paris, he noticed the interest Helena showed for Max, but he didn't really understand it, since to him, Max had seemed contrary and disbelieving. And the truth was that he hadn't liked him at all; he was arrogant and stubborn. Diego had that encounter in his mind for days now.

"You know what, Helena, I never asked you about Max. It's weird, but for some reason I've been thinking about him lately. Who is he? Why did he want to see you after so many years?"

"He was a classmate of mine when I was studying at the University. It seems that he's having some dreams that are really troubling him. You know? Reasons from his past are presenting themselves to him so that he can resolve them."

"Well I saw that he treated you with a lot of familiarity. But I didn't like it when he got stubborn and aggressive. Was he always like that?" He was surprised to hear that remark leave his mouth and he realized that he was jealous of Max.

She perceived the emotions in his question, but just responded, "No, in fact he was really nice and a lot of fun." Diego felt even more annoyed, but fought to control his feelings. Helena continued. "Look, it's not going to be at all easy for him to confront his situation. What he doesn't know yet is that it was he who asked for this opportunity before being born, even though he doesn't remember it. And in this there is no going back."

"Well I only saw that he wasn't accepting anything you told him. He's so full of himself and he's really into playing the role of the great scientist," Diego insisted.

"Did you wipe your feet before coming in? Or did you leave mud all over the floor like the other day?" Diego jumped up out of his seat and searched the floor to see if he had dirtied it; when he saw that it was clean, he turned to look at Helena, and by the twinkle in her eye he realized that she was saying something else. With this scare

Diego put more of his attention on what was transcendent. "We're here to cleanse ourselves through study. That's why it's vital that Max comprehends that this is his opportunity; he's approaching his destiny. To each person there comes an opportunity to face something and advance, Diego, but the hard part of this is knowing how to recognize it in the moment. These possibilities come in specific moments and we shouldn't let them pass, in spite of the difficulties that they may bring us. Growth implies an effort, doesn't it? The situations through which an Ego needs to pass may be for different reasons."

She explained in more detail. "There are the reasons of karma, whether it be of the individual or of the group to which he belongs. There are also the reasons that an Ego can choose in order to promote its own evolution. But above all of these is destiny, which leads an Ego to its Realization within the Kosmic reasons. Look, let's read this study so that you can understand a little more about destiny and the reasons that fall to us in life." She opened a notebook of studies and read to him: ***"All of the Egos have 'experiences' which are the traffic of***

reasons that can change, variants. They are acquired in the spiritual planes (of the Trebolo) *and are human prefabrications."* To explain, she commented, "This means that one decides certain circumstances for oneself. For example, when one decides to have children or not, or be in a specific family in order to pay a debt. If an Ego had too many material reasons in one life, it may program in the Trebolo that it have an austere life, since it understood that this path distracted it from what was important. Sometimes it's allowed that an Ego in the Trebolo chooses circumstances that are going to help it in its growth; even though, in the Ansibir the Ego is not konscious of having chosen them and at the moment of living them, it rejects them because they seem contradictory."

She continued reading the study. *"The 'reasonings' are Astral, they are causes that are established in group, for example, reasons programmed by the configuration of the planets, the horoscope. It is the synchrony with the Astral world [...], the relationship of the individual with the Kosmo that he inhabits and they are gifts, graces that are granted even though we do not always understand them as such when we consider them adverse. They are not understood."* She commented, "We can understand this as the causes that mark a generation or the fact of having been born in a specific time or having certain graces or limitations. These reasons sometimes cause us pride or suffering because we don't understand their purpose. But we are all born with what corresponds to us so that we can advance to the next stage of our development."

Helena continued with the study. *"'Destiny' are the reasons that prevail in the Kosmic strata and beyond the Reasons of Life. [...] Destiny is a vital cause, like a spiral of force that drags along, absorbs and distances the other causes. That is why he who is touched by destiny lacks in relation to the other experiences: loves, economic success, etcetera.* We can think of artists," Helena remarked, "like Beethoven o Gauguin, who could only develop themselves in what corresponded to them, and suffered various deficiencies in other parts of their lives." She returned to the study. "But in the majority of men *the experiences and the reasonings predominate and, in a minimum, destiny. As an Ego has a greater elevation, it is subjected more to the Kosmic reasons, to its destiny.*" Helena clarified, "We can see this in the case of a person that has had a spiritual search throughout various incarnations. He no longer finds a place in the material world because he senses that he will not find his truths there. The things that other men search for, such as love, money, family, fame, etcetera, no longer fill him. The spiritual elevation that he has acquired throughout his lifetimes, distances this person from the everyday reasons so that he can approach the Kosmic reasons. We can appreciate this for example in the case of someone who understands that they should separate themselves from what their life has been and from their family to go live in solitude in a half-forgotten village because they know that in this way they will be able to advance and develop; even though, for others this is a senseless act. *When destiny arrives, the person begins to question himself, he loses his balance in the face of what he lives.*

"However, the majority of men *live predominantly ruled by the spiritual experiences, commitments that fill their lives with what they will later understand to be an absurdity. The configuration of* *these acts may be rewards or reprimands.* These spiritual experiences are made in the planes of Trebolo and are commitments that an Ego makes to resolve a karmic debt or to overcome a deficiency. Or in the case of an Ego that made an effort in its previous life, it may receive faculties or graces that will facilitate its growth in the next Ansibir.

The Reason Life has as its basic meaning an apprenticeship, but of the lessons we accept some and reject others without understanding that they are given to us for a study. Like a soldier who suffers a training that exhausts him but in battle comprehends its usefulness. When an Ego understands the why of the circumstances of its life, it facilitates its experiences and it feels joy at knowing that it is advancing. But if not, it suffers because of the adjustment that it has to live. *The apprenticeship implies an effort and the effort implies sadness, pain and suffering, when it is not understood. The harmony and the understanding give the joy in the apprenticeship (17 Jul. 89).*

That's why I tell you that Max is finding his destiny, because this is his moment to face the Kosmic reasons of his evolution. He doesn't understand it, but that's why his life is falling apart before his eyes, and the commitments that he has in the world will pass into a secondary importance in order to help him fulfill the Kosmic reasons. These take away importance from the spiritual experiences (commitments made in the Trebolo), since they are only human schemes. For example, if you had to live with a family in order to fulfill your karma, but in that moment you encounter the Maestro or a door to be able to understand the Konocimiento, all of the other reasons that you had to complete are no longer valid. The kosmic reasons of your destiny then take the importance.

Max is living a battle inside himself because the Ego is konscious that it has to resolve the unsettled issues that it has been dragging along for centuries. Of course he's going to defend his world aggressively, but Max is an evolved Ego; that's to say, in another time he achieved a greater level of comprehension and development, he had a great responsibility, and hopefully he'll understand it. Sure, he's made a lot of mistakes along the way that have confused him and sidetracked him." When she saw the expression of surprise on Diego's face, she said, "And who hasn't? It won't be easy for him; the motives of his quest are very involved with power and dominion. But as they say around here, 'you reap what you sow.' You know, Diego? The Konceptos never place us in a situation that we aren't capable of overcoming."

More than three months had gone by since Max, perturbed and angry, had bid goodbye to Helena in the subway station in Paris. After having left her that day, he had continued walking to his job in the Institute. A torrent of ideas had flooded his mind. 'How is it possible that Helena said all those things to me? Who does she think she is? What irritated me the most was when she spoke of an unresolved commitment. What the devil is she talking about? This time she went too far. Talking to me about things that happened thousands of years ago as if it were yesterday. I don't know how she expects me to be able to remember it. I can't even remember Michel's birthday.' His thoughts wandered to his ten-year old son. 'He always complains about that, something which of course his mother never hesitates to remind me. Oh, it's a good thing that Yvette and I got divorced. How that woman nagged me... At least Helena isn't superficial like her, although she goes totally to the other extreme.'

Now, his mind returned once again to that meeting with Helena. 'Many of the things that she told me then seem unbelievable. Even though Helena never liked fantasizing. Some of her ideas were quite interesting. It really would be incredible to be able to remember the entire sequence of our lifetimes. Umm, so many possibilities... ' He had thought of calling her several times, but always stopped himself, intuiting that if he did, a huge change would come. When he began to think seriously of what Helena had told him, he realized that he was going to have to reformulate his thoughts about life and decided to banish those ideas from his mind.

He had made an effort to incorporate himself into his habitual life, but was unable to. Before, he had considered his situation as just a bad spell, but it still hadn't ended. After having seen Helena, things worsened rapidly. Because he had been so distracted and distressed, he made a critical mistake in his laboratory and now was facing a lawsuit against him. The director of the Institute was very irritated with Max, as he had noticed his mental lapses and he warned him that he'd have to find a way to pull himself together and focus on his work. The Institute had made a very costly investment in his investigations but they hadn't produced any economic results after three years. His personal life was also turned upside down. Yvette his ex-wife was pressuring him constantly because she needed more money for their son Michel's expenses and activities. The dreams continued and each time they perturbed him more, preventing him from getting enough sleep. He felt irate and anxious and didn't see an end in sight for his misfortunes. None of his colleagues had been able to help him and he knew that he couldn't keep taking tranquilizers any longer, or his work would be even more affected.

That day, to top it all off, the director of the Institute had scheduled an appointment with him to talk about "something important" and Max felt a knot in his stomach. He supposed that it had to do with the pending lawsuit. In the midst of this riot of emotions, Helena's words remained in his mind, repeating themselves with a most significant resonance. He thought, 'how odd that all of my problems were accelerated after having talked to her. Well, of course, I was distracted by all of the strange things that she told me. But if everything she told me were true? No, it can't be.' Tired of not being able to reconcile himself, he decided to concentrate on his meeting with the director.

It was five in the afternoon when Max entered the office of his department chief, Doctor Vincent Tellier. The serious look on his boss's face made the little bit of tranquility that he had achieved evaporate within seconds. They greeted each other cordially and Max sat facing the large and imposing desk. Max tried to calm his thoughts and focus his mind on the meeting.

Vincent leaned forward over the desk, resting on his forearms and said in a strangely formal voice, "Well, Max, we've worked together for a long time. Your efforts over the years have been a great help in the Institute's projects, but you know that everything depends on results. And at the end of the day, we're like any other company that measures success according to earnings, which up to this point we haven't seen in your projects." He cleared his throat and leaned back in his chair. "Really, it makes me very sad that it's come to this, Max, but we have to reconsider your position here. We're grateful for all that you have contributed over the years, but we feel that it would be best for all to terminate our relationship at this point." He explained how they had arrived at the decision but Max didn't hear the rest of his words.

He left the office in a daze, as if he were inside of a nightmare and unable to wake up. He couldn't believe that after fifteen years of work and dedication they would have fired him like that. With a stir of feelings in which anguish and anger stood out, he asked himself, 'how did they dare to do this to me? To me, who's put in so much dedication and effort. To me, who's left my whole life in their laboratory.' He passed by his colleagues who were in the halls. They stared at him, wondering what could have happened to him. But he didn't even realize they were there. He was trembling when he took his jacket from the office. He went out onto the street and decided to walk and try to calm down enough to be able to think.

'What am I going to do now? I'm 41 years old; it won't be easy to get another position like this one. How am I going to explain this to Yvette? The first thing she'll ask me is how I'm planning to support them. And Michel's school…It's all because I was acting like an idiot, paying attention to some dreams and thinking about the things Helena told me.' He didn't know how long he had been walking around the city when he noticed that he had arrived at the Palais-Royal, where he had his penultimate encounter with Helena in February. He was remembering the conversation that they had there, when he revealed his pain to her at having lost his sister Isabel. When he thought of his sister and of the serenity and acceptance she had before dying, he felt something break inside of him. He entered a cafe, sat at a table and ordered a whiskey to mourn the loss of his life which, for some strange coincidence, had finished falling apart after having seen Helena.

Helena awoke early, feeling uneasy. She got up and walked downstairs to turn on the coffeemaker. Sitting herself at the old table, she asked through the Kommunication about what was bothering her. She understood now that Max had arrived at a critical point in his life and that his mental and emotional problems were distressing him excessively. Now she knew the reason for those troubling images that she had seen in her dreams lately. Max was projecting his agitation and rage and blaming her for what had happened to him.

She remembered the lesson she had studied: *'The lack of balance of a person affects, echoes in their environment; for example, a state of ire. Even more damage is caused by a distressed person who has awakened in himself more abilities than the rest, if they don't control them. […]In Life, everything is interrelated, that means that everything is one and the same' (7 Jan. 89).*

What she understood upon asking left her forewarned and intrigued; she knew that Max would come to look for her. She'd have to prepare herself for this meeting. 'This is going to be very interesting,' she thought.

When Diego arrived that night to study, Helena asked him, "How have you felt? Have you dreamed or seen anything in particular?"

Diego contemplated her remark and said, "Umm, let me see… well, thinking about it, no Helena, I've just felt uneasy." Diego hadn't given it any importance, but now he perceived something in Helena's words, and he questioned her, "Why do you ask? Has something happened to you?"

"You know what? Something's going on with Max. Call him and see what he tells you," Helena instructed him.

"Me?" It was the last thing Diego wanted to do, but he agreed to his Maestra's request, saying, "Ok, Helena, if you want."

"Invite him to come and spend a few days in Remoulins. It seems to me that he needs a refuge and a friend."

'Uh-oh, what's going on here?' thought Diego. They went into the living room and even though he was skeptical, he dialed Max.

Max was seated at his desk in his apartment in Paris making calls, looking for work. When he hung up the phone, it rang immediately with Diego's call.

"Hi, Max, it's Diego."

Confused, he replied, "Diego? I don't know any Diego. Ah, yes, of course, you're Helena's young friend. How are you? How is your father doing?"

Diego felt a contraction in his abdomen when he heard the question. "Well, the fact is that he died a little while ago, but thanks for your interest."

Max felt badly for not having thought before asking the question. Trying to understand the motive for the call, he remarked, "My condolences, young man. But I hope there's no other problem there."

"Oh, no, everything's fine here. I just wanted to say hi and see how you were. And well, I thought that if you can escape from your job, you could come round and visit us. Maybe a change of scene would do you good. It's really nice here in the south."

Max swallowed the discomfort that Diego's words produced and decided to hide the truth of his situation. "Well, I don't know if you know this Diego, but life in the laboratory is very demanding. Right now I don't have time." He still didn't understand the objective of his call. He remembered his parting with Helena in the subway and feeling remorse for the abrupt way that he had left, added, "I appreciate your interest and ask that you say hi to Helena for me, I hope she's well." He was going to say something more, but decided to remain silent. With that, he hung up the phone, feeling jealous and thinking, 'is Helena with him now?'

Helena enjoyed the faces Diego made as he tried to appear interested in Max's life. Curious to know how Max had reacted, she asked Diego, "And did he agree to come?"

Diego recounted the conversation to her. She smiled and remarked, "In life, as in the Trebolo (death), we only see what we want to see. Max still doesn't want to accept what he's living, even though the Lux is right in front of his eyes. I hope that he doesn't go blind."

Another two weeks passed and Max was unable to pull himself out of his depression. He hadn't been able to find work and the date of the meeting to resolve the lawsuit was approaching. He was afraid and angry, full of resentment. He didn't understand why his life had turned upside down and he hadn't been able to do anything to correct it. It seemed to him that all the doors were closing before him and his desperation grew as time went by. That night he went to sleep thinking of what he would do if he really couldn't fix his life.

In the morning he awoke hearing the beating of his heart resounding in his temples. He felt as if his head was going to explode, the pain was that strong. He had had another terrifying dream, and that was the last straw. He knew what he had to do. He sprung out of bed and packed a small suitcase with changes of clothes for a couple of days. He left his apartment and took a taxi to the Gare de Lyon station and got aboard the fast train, heading for Avignon.

At midday, Helena was working in her garden when she received a phone call. It was Max advising her that he was in the train station in Avignon. He said that he needed to speak with her urgently. She explained to him how to get to the house and sat to smoke a cigarette on the back patio while she awaited his arrival. She analyzed Max's situation and asked herself, 'will he be tired of fighting against his destiny yet?'

A short while later she heard a taxi pull up to the entrance of the house. She saw Max getting out of the cab with his suitcase in hand. His aspect was quite disturbing; he looked untidy with a two-day growth of beard and dark circles under his eyes that indicated a prolonged bout of insomnia. Helena went out to receive him, curious to hear what had pushed him to take this step of visiting her so impulsively.

"Did you come to take a few days of vacation, Maximilien? I hope that you brought your swimsuit. The weather has been really lovely. Of course, if I remember well, a little bit of cold didn't stop you when you stripped naked and jumped into the fountain at the Place de la Concorde that flamboyant night."

Her words, so unexpected and out of context, disarmed Max and his distress disappeared. No one had called him Maximilien since the years in the University. He blushed at remembering that scandalous incident of his youth and relaxed a bit. He replied, "But what a shame, I didn't know that you had a pool, Helena. Or perhaps there are fountains in this god-forsaken place?" He left his bag on the ground and greeted her.

189

"Alright, tell me, what happened to you? Why the rush to come to see me?"

Max vacillated. His good humor evaporated rapidly and he became serious again at recalling the reason for his visit and the dilemma he had. On the one hand, he had come to ask Helena for her help. And on the other, he blamed her for everything that he had lived in these three months. He tried to reconcile the two parts and said, "Ok, I came because my life is a disaster. The truth is that I've lost almost everything since I saw you." In spite of his efforts to control his emotions, he started to feel angry because of all of the changes that he had lived and which she represented. "It's that the things you said to me affected me deeply and for some reason I haven't been able to stop thinking about them. That's why I was so distracted that..."

"Ah, you blame me for the situation in your life and you came to ask me to help you to get it back?"

"No, no, it's not like that, Helena. Well, I hope that you can forgive me for having been so rude that last time that we saw each other. I know that I made a mistake." He stopped and began again, trying to express what he felt. "It's just that...well, the truth is that I came because you're the only person that can help me." With these words Max overcame his pride and with this something inside of him opened and he could express himself more sincerely. He told Helena about all that had happened in the months since that meeting and of the dreams that continued to torment him.

"Look, if you only want to feel better, go to one of your colleagues and have them give you a magic pill. If you want to recover what you lost, get a job in a pharmaceutical company or invent a cream that removes wrinkles. Or go travel the world; maybe after three years or so, you'll forget about your troubles and the dreams. Or maybe not. I'm not very sure."

Max became livid again in the face of Helena's words. "It's always the same with you; you don't take me seriously and you mock me. I'm asking you for help; don't you know how difficult it is to reach the point of having nothing?" He ran his fingers through his greying hair, trying to calm himself and breathed to focus his thoughts. "I'm asking you to help me to understand what's happening to me. Will you do it?"

"Are you going to accept what I tell you this time?"

Max was surprised to hear the strength in Helena's voice and swallowing his pride, replied, "Yes." He felt fear upon pronouncing that word, but in spite of it, he asked, "Just tell me what I have to do so that all of this fixes itself."

"First, we can't stay here outside all day. Let's go drink something to refresh ourselves."

When they were seated at the table in the kitchen, Helena served him an iced tea. When she saw that Max had calmed down, she said to him, "Alright, why do you think this is happening to you?"

"I don't understand it, I only know that from having been up here," he signaled with a movement of his arm, "now I'm here," he said, marking a lower point. "I don't know what I did to deserve everything that's been happening to me."

"There's a reason behind this. When it falls to an Ego to face its reasons, the opportunity to advance arrives, but first he needs to detach himself from everything that blocks his way. This is what you have been living. You're approaching your destiny, and for that reason you have to separate yourself from all that impedes you from Realizing yourself and from all that can sidetrack you. It's a moment of rupture, it's like a birth. Of course, in your case you're so hardheaded that the blow was colossal, but it's just to put you in a situation which shakes you up so that you can perceive something different. I imagine that you feel as if you've been kicked off a train that was going full speed, right?" Max laughed upon imaging the scene, but it was due more to his being nervous. Helena continued. "Look, *in the material, man should know that everything is for an apprenticeship, what comes to him is for this purpose and he should accept it. If he enters into rebellion the road becomes more difficult. Everything arrives in the precise moment. We should have confidence in that the Koncepto conditions our life and the delays are necessary in order to prepare important changes" (11 Jun. 88).*

Max made an effort to see it that way, trying not to impose his ideas. "Ok, it could be like that, I suppose... Then everything that has been happening to me has a reason to be."

"Well, yes, Max, but as you have spent many lifetimes without wanting to see or fulfill your reasons, they're pressuring you so you let go and see what's really important."

Max leaned back at hearing that and contemplated this idea. "Yes, as you say, I realized something last night and I believe that now I understand why you indicated that my dreams were remembrances of the past. When you said it in Paris, I thought it was impossible, that you were fantasizing, but now I believe you."

"And what made you change your mind?"

"It's that last night I dreamed something which seemed so real and affected me so deeply that I decided to get on the train and came here." He paused, thinking of the distress that he had had in the morning. "I dreamed about the research I did in the Institute and suddenly I saw myself in other, distant times; I believe they were images from other lifetimes, and in several of them I was doing almost the same thing I do now. I saw that there was a pattern that repeated itself constantly, like you told me. I felt the excitement and the curiosity that have led me so many times to investigate, to explore, to try to understand how the body functions. But I realized I wasn't just trying to understand it, but that I have tried to modify it or 'improve it', shall we say. This has been the intention that has linked my lives. I was able to glimpse some of the consequences of it: matter has gone on transforming itself, but it has suffered more of a limitation than an improvement. It was very difficult to accept this. Was it really like that?"

"Yes, in earlier civilizations, like Atlantis, Lemuria, Mu, and others previous to these, that we no longer remember, or in some more recent ones like the Mayan or Egyptian, they modified the gene, matter and konsciousness. The civilizations previous to the Atlanteans received the teachings of the First HTime and with them achieved a completeness. But there came a moment in which, confused, they decided they could act for themselves without asking permission and they distanced themselves from the guidance of the Konceptos. What they did was a repetition of the first error in the Kosmo of the Trinak: that of wanting to encompass more than what corresponded to them. With their actions they disrupted the Laws of their development and had to be removed. Each civilization that followed the first races had the memory of these initial teachings. But in their minds these lessons were no longer complete because of the limitation that those men suffered when they used them wrongly. In time, they tried to repeat the 'achievements' reached in the past, without accepting that now they had new lessons to learn. And each time the same thing occurred: these civilizations were erased from the face of the Earth so they wouldn't cause any more damage. With each destruction, men suffered a limitation of konsciousness and of the abilities they had. When we arrive at the era of the Mayas or the Egyptians, the memories of these first teachings were already very disturbed and distorted, and these races once again committed great violations. But what has been the goal of all of this? There are three intentions that we should recognize.

[...]We should reject a negative intention, which is that of dominating one another, or of forming a supreme Hueste (Host) that dominates the rest, as it is currently doing.

Another of the great errors is to conceive of an eternal matter, even though the Ego does not really know what the Eternal is nor does it understand it.

Another of the negative reasons of transcendence is the endeavor of wanting to conquer the Universe or Fusion (25 Oct. 11). The intention of this last reason is to leave the Planicio Terráqueo before concluding the apprenticeship, but this is not possible, since we are subject to the Laws that the Konceptos mark for us.

Everything that happens currently can be reduced to these three errors. Nothing that man does is new; it's simply a repetition of something that was already done and that didn't lead him to any positive end. But now we have the possibility of changing our actions and avoiding the coming of another reversal in the konsciousness of man. We should forget about the practices of previous cultures and civilizations because they're not in the exact HTime anymore. The Kosmic teachings of the Third HTime are new and they can take us out of this constant repetition which is the karma in which we have been enveloped for millions of years, in order to project ourselves toward new and more complete reasons."

"That's why I came to see you, because I thought that with the studies you have, I can find a way to 'correct' what I did wrong in the past without causing any more damage. Ok, what I want is that you help me to clean up all of this."

"Max, things aren't like you imagine." She got up from her seat and went to look for her cigarettes. She offered one to Max and continued explaining. "Look, what you're telling me now is that you want to learn so that, according to you, you can do what you consider to be a good thing. But how can you know if it's good or not if you don't know the purpose for which it was realized? Do you see that when you think like that, you continue to impose your desires, Max? What you have to accept is that you should let yourself be guided by the Konceptos that govern us and have a total submission to their reasons. *Sub-mission* is fulfilling the *mission* that each of us has in the world, within the reasons of the Real, not in the fantasy of the material. To arrive at this you should put an end to your desire to impose and to change everything."

Max leaned forward, resting his arms on his knees. Helena noted various things in him: his conflicts, that instinct that propelled him to understand, and his desire to resolve the misfortunes of his life.

She remarked, "It's good that you want to comprehend, but you should understand that it's not to undo what you did. The damage is done, to you and to the rest. What we have to do is try to see, analyze, and cleanse these erroneous reasons in us, understanding the motive that led us to participate in them so we don't repeat them again. *We are in a process of finding ourselves again with the perfection of the beginning, it is a process of cleaning [...].When we understand, one is cleansed, is purified; the Ego by mandate, not in rebellion, will find that relationship with the animal, with the stone, and with the experiences in the Kosmo of the Trinak (8 Oct. 88).* If you decide to approach the Konocimiento it should be in order to understand the reason for which it has been given to us and carry out with It what is indicated to us. It is not given to us to modify or improve things, we aren't creators; on the contrary, it is given to us as a means of understanding ourselves and surmounting our deficiencies.

You know, we have to accept that we don't know everything. If we follow our own instincts they'll only lead us to repeat what we've done wrong for so many centuries. All of our problems here in the Planicio Terráqueo have their origin in the fact that we haven't accepted the rules and the conditions of the lessons that we have to live. We don't ask permission before acting; we impose what we want because of arrogance. We don't accept our limitations, nor the fact that not everything is permitted to us. And these impositions have had many unexpected consequences that keep on affecting us in ways we're unconscious of. Look, in the studies I learned that *the Egos in the First HTime wanted to change the structure that had been conceived for this primary stage and they filled themselves with complications because of their impertinence of modifying that which was well planned; in this was manifested the reason of the first rebellion: I can get ahead, without consulting (21 Oct. 88).*

It should be remembered that Ego means that which develops itself in principle and forms part of the Voluntad Suprema (Supreme Will) (Carpeta Dorada).

"But how is it that something that happened so long ago can continue to affect us?"

"Ok, I'll give you a simple example. Upon not accepting the Voluntad and breaking certain laws that were delivered to us in the beginning, we caused a huge damage to ourselves. *Man, and his human structure, was formed to survive with little food, and with this it was brought about that he be exempt of the whirlwind of feelings and confusions, in order to search within himself and encounter as a fundamental principle, harmony, and the contemplation in the environment created for him and by him. This environment should feed him in order to find the truth hidden within the reasons of the Apparent, not to create a maximum apparent, which is what currently, man has done.*

The ingestion of excessive food created an imposition in the matter, which was centered in the possession of the environment, that is, the hunger was not only for food, but hunger for the environment itself, in its different manifestations. With more food, greater desire for power. Power is taking over the environment. All this brought as a consequence, an alteration in the Espíritu Huella[...] (30 May 96). Upon altering matter, and as a result, the Espíritu Huella, we entered into a maladjustment in relation to the reasons of the Planicio Terráqueo and the Universe. We can no longer perceive them fully. We fell behind in respect to the Time of the Planicio Terráqueo and we entered into a repetitive one. As a consequence, our matter suffered repetitive distortions that we call illness. You see, then, how the consequences of an act that had no reason to be multiply themselves and echo in everything."

"Wow, yes, it's very complex," exclaimed Max.

"In your case, Max, it's not your desire to understand the world that's wrong, rather it's the intention behind your search. All of our errors can be reduced to wanting to have more than what corresponds to us. As I told you in Paris, everything we do and think remains registered in the Espíritu Huella. And it's vital that we understand the reasons of our past so that we can advance. If we don't arrive at having clarity about what we have lived and about what motivates and drives our actions, we'll keep repeating the same thing over and over, as you already saw in your dream."

Max was finally seeing his smallness in the face of the work that Helena was describing to him. He felt an emptiness inside and a little bit of fear at contemplating the implications of Helena's words. He asked her, "And what would I have to do to change this, Helena, if as you say, it's the culmination of centuries of acts that shouldn't have been?

"A pequeño, or little one as the Konceptos call us, can as a first step, have the sincere desire to understand the reasons of his past and put his efforts into this. But in order to understand them fully, he needs the guidance of and Kommunication with the Konceptos. This comes only by way of an initiation into the reasons of the Konocimiento."

"Can you help me, Helena?"

"Yes, Max, but in this way. Stay here in Remoulins for a while and we can talk. I'll tell you about the studies and you decide if there's something in them that can help you resolve your problems. Deal?"

"*Ça va*, Helena."

Chapter 11
Attraction

Why do we fight?
Why does everyone want the land or
the possessions of their neighbor?
Envy is foolish.
NONSENSE

The three days that Max had thought to stay in Remoulins turned into weeks. During these days Helena had many conversations with him. In this time, Max used his great persistence and concentration to strive to understand the studies that Helena explained to him, even though he had to make an effort to open his mind to ideas that didn't fit in with what he had learned before.

Helena advised him that he keep busy in his free time with a physical activity, so that he have a balance. The first thing that occurred to Max was to look for work in the hospital in Avignon, since it was the only thing that was familiar to him. But Helena indicated to him that it would be better to do something totally different in his life, and she suggested that he help Diego in the hardware store, knowing that this change would help Max confront his arrogance, and in this way to see his own reasons. At first, Max had felt belittled and saw this job as beneath him. But as the days passed, he realized that it helped him to be occupied in physical tasks and not just mental ones. He even started to enjoy the interaction that he had with the townspeople, although it was very different from the deference that his patients had shown him before.

To his surprise, the comfort and closeness that he had felt years earlier with Helena emerged again, and he realized that he felt a great attraction for her. Her agile mind and her sarcasm had always appealed to him, but there was something more interesting in her now that attracted his attention and made him want to be close to her.

Due to the obvious feelings of Max towards Helena, the atmosphere in the hardware store became tense when a small rivalry was born between Max and Diego in regard to Helena. Diego also felt a pull towards his Maestra, even though she was almost twenty years older than he. He was annoyed by the way Max spoke to her; they both had very quick minds and there was an affinity in their way of thinking. For his part, Max resented the closeness Diego shared with Helena because of the studies. Max demonstrated his jealousy in the comments he made to Diego, belittling his ideas, pointing out that he was just a kid, or flouting his own intelligence. The situation worsened until one day Helena called them to a meeting at the house in order to discuss the issue.

"Alright, we're going to resolve this now, because if it keeps on like this, it can get out of hand. Look, it's obvious that you both feel an attraction for me." Diego blushed and got nervous. Max began to play with the pack of cigarettes in his hand. The two men started to deny it, since they saw in what direction the talk was headed. Helena, without feeling the least bit uncomfortable, continued.

"Yes, I feel what's happening and I see the friction between you two because of it. These energies are very strong and when you fling them around between you, you unbalance us and additionally, sidetrack Diego from our goal, which is study. What you think of as an amorous attraction for me is really something else. You're attracted by the Lux and the energy that are in me since I'm initiated, and because of the elevation that I have. When an Ego studies and understands through the Konocimiento, he has a relationship with the Kosmic reasons and with the Konceptos instead of submerging himself in the Continuo (Continuum) of human reasons and thoughts. This is elevation, which only a few appreciate. It's a greater energy, since the initiated Ego receives a greater charge of Lux. Those who are around him feel it and they seek it, trying to connect themselves to it. All of this is related to the structure that the Egos had in the Kosmo of the

Trinak. And it happens that way in all human relationships in terms of physical attraction or love, it's just a play of energies.

Ok, in the Kosmo of the Trinak, the Egos were related in a hierarchical way, which we can visualize as a triangle. Within this structure there were many levels of evolution and konsciousness, with the Egos of greater evolution in the upper part. The responsibility of the Egos of greater development was to teach the less developed Egos so that all would rise through the different levels of comprehension. It was like a chain and in every 'stratum,' an Ego was in relation with Egos less konscious than itself in the lower adjacent level and with Egos more konscious than itself in the upper adjacent level. Well, this function continues among the Egos in the Planicio Terráqueo. Each Ego should receive teachings from Egos in the upper stratum and in return, teach those in the lower level. This action of giving and receiving is part of the makeup of the Yo Ego Deidad. It's a give and take of energies between Egos in order to balance themselves and complete their function. In part, for this reason we have the idea of seeking a complement in a partner.

When an Ego relates itself with a person who is not at its level of development, and the Ego that is less prepared has a greater charge of energy (or if there is a bond between the two), this can pull the more developed one 'downwards,' causing the Ego of greater development to end up involving itself in experiences that are not in accordance with its level; then this Ego 'devalues' itself. For this reason certain relationships between two people can be very destructive, or in other cases it's the opposite and they impel an Ego to advance.

It's all about energy and how it flows between Egos. One Ego feels an attraction towards another Ego when this one has a greater charge of energy or when there's an affinity in their sonority and in their reasons. We all know of the famous cases of ugly men who are like honey for women. It's the man's energy that attracts them. The energy that a man receives upon attracting various women, in most cases, is used to attract even more."

"Ah, so that's what happened to my ex and the ugly toad she took up with. And she says he's her prince," said Max, letting out a guffaw.

"The attraction and relationship between two Egos should be temporary so that it serves as a complement and as a lesson. It shouldn't be something fixed or permanent, except in some very specific cases. When there is a balance between two Egos, there's harmony. Their energy is equal, and it's when we say that we're in love. But when one starts to take more energy than the other, the balance is broken and that's when the two should separate. For that reason the custom of forming a permanent relationship between a couple causes problems and damaged for both. There are moments in which the Ego can be involved in a relationship, but there are others in which it has to walk alone, and these bonds and energetic commitments normally prevent it from doing this."

"And love, Helena, where does it fit into all of this? Because some believe that love is everything, that it's the maximum in the Universe," commented Diego, relieved by the explanation that Helena was giving them.

"Ah, Diego, well true love is beyond human reasoning, it's the impulse that makes us approach our beginning. *Love, for the Konceptos, is the exaltation of Truth itself (28 Nov. 87)*. What men call love is as I said, a question of energy. As it says in the studies: *We are told of the 'ideal couple' and other fantasies. We should take love as a harmonization of our matter, and there will be periods in which it is not given because it is necessary to be so. We should not see the one who arrives as the 'hope' of which a false programming speaks, it is a harmonization and is ephemeral (but should not be rejected because of this).*

A stable relationship is a merit, and is for special reasons. Our life in the spiritual is development; in the material, acceptance; and in love the motivation of harmony (11 Jun. 88).

"We've talked of how the attraction between Egos functions by means of sonorous affinity, now I'll explain it according to HTiempo (HTime). *Attuning oneself with someone is to enter into the Time of that person, that is the secret of love, which happens when two people are in the same Time. If there is an affinity, a Time is shared, but when one of the two changes Time, the parts fall out of tune and disputes emerge. If there is attraction: the same Time. If there is repulsion: different Time. The bodies are attracted to each*

other in relation to their energy and (part of) this arranges itself in Time.

In the future humanity there will not be stable relationships but encounters in which two pequeños unite, understand and separate. At the end of the HTimes, all of us will be attuned in one same Time" *(6 Feb. 88).*

"Then is the complementation of energies the reason that we exist as men and women, Helena?" Max asked, trying to change the subject.

"Well, the Ego in itself has no gender. This apparent duality of the matter is part of the condition of the Ego in the Planicio Terráqueo. *[…]Upon entering the Planicio Terráqueo it was divided in two in order to give it the characteristic that we call feminine and masculine gender; this is a fantasy since the chemical composition of one is equal to the chemical composition of the other, it is simply a difference between having a higher amount of one hormone or a lesser amount of that hormone"* *(11 Apr. 08).*

Max wanted to know more and asked, "Then, we can forget about pheromones?"

"There we go, Max. In the same way that the Ego seeks an energetic complementation, matter has its own search for complementation with other matters and with the environment. It's what we call feeding, which doesn't just refer to the act of eating. If we examine the energetic relationships between Egos in matter, we see that *man would be an Oval-shaped Mass that feeds, rejects and discards, but not just food through its mouth and the waste of excrement, rather its own energy and that of the environment, and the man-woman human relationship is a clear example of it (it is called love). It is fed from chemical and physical substances between themselves and their environment. The same happens with the plants, the same with an animal (example: dog, bee, butterfly), the energy eats another animal and another animal eats him (11 Apr. 08).* This is the true food chain that is acting on all levels of our interactions. That's why we say that we live in a Planicio of multiplication."

The tension between Max and Diego relaxed due to the explanation that Helena had given them. Diego understood that what he really felt for Helena was admiration and an attraction for what she taught.

Once the chat was over, Max excused himself and left, leaving Diego in his class with Helena. Diego took advantage of the chance to ask Helena more about the relationship between Egos.

"You know, Helena, I remember that when you told me about the layers of the Ego, you said that matter is made of triangles. Does this have something to do with the way we relate to each other?"

"Yes, Diego, there is a connection. This is said about matter because we live within the triangular reasons of this Kosmo, and our matter was formed according to these Laws. The triangular Akt is a way in which reasons within Life develop themselves; there are other geometric Akts that lead to other ends which we can't understand. *The triangle manifests two reasons which in one moment are divergent between themselves, in order to later join themselves in one alone. The Impulso Impherante (Ruling Impulse) projects a yes and a no before its dominion, to later consolidate a harmony that subdues all (5 Aug. 97).* This apparent struggle between two contrary reasons should lead us to an elevation and a harmony. Regarding the reasons of matter, *every body or grouping has a mass that converges and a waste which is divergent. This is what we call the separation of the triangulation. All matter is subject to this reason (2 Dec. 93).* Because of the separation, matter suffers deterioration.

Matter always relates itself to the environment, or with other matter, based on triangulation, searching for a complementation and a point of reference. *There exist triangular particles that are of the Planicio Terráqueo itself, and they affect us, entering and leaving our own casing. This entrance and exit coincides with the actions of our matter, such as eating, sweating, defecating, etcetera.* In these triangular particles *each part of above or below, according to its power, will search for the upper or lower complementary part, in another particle. Each triangular particle insists on its reference (27 Nov. 93).* This is what we perceive as an affinity or a battle between matters and the environment. It's the same as what we see in the complementation between Egos.

The Ego also has a triangular makeup; it lives an apparent duality between two divergent reasons (what we call positive and negative) to arrive at a Real harmony, called Armonio, which is a point of convergence or surmounting. *The Ego has to find its perfection; the*

perfection of the Ego is comprehension and the action in the positive, negative *and Armonio in plenitude" (28 Aug. 90).*

Max went to his room in the boarding house recalling the talk. He thought about the attraction he felt for Helena since his youth and accepted that her unusual way of seeing the world was what had always appealed to him. But he couldn't keep from admitting that her green eyes, and all of her, for a matter of fact, also attracted him.

He understood what Helena had just explained, but he was afraid to think of what it meant for him and for his life. If the attraction he felt for her really was for what she represented and for the studies that she taught and not for her, then what did that mean for him? He knew that if he committed himself to that world he'd have to leave the other behind. These thoughts generated a huge conflict within him.

In his last days in Remoulins, Max spent his time alone, walking in the countryside to think and reflect on his life. The moment arrived when he had to return to Paris to resolve the lawsuit against him, so he went to Helena's house to say goodbye.

He was surprised to hear Helena's reaction when she said, "Don't worry, Max, you'll see that everything will turn out alright. You'll even get job offers. Your old colleagues are going to help you."

Max felt he still didn't want to commit himself and tried to maintain a foot in both worlds.

"You know, I'm not so sure that's what I want, Helena. Maybe I can come back here in a few months and we can talk some more about the studies. Or maybe you can send me some when I return to Paris. I can study on my own."

"Yeah, sure, I can send them to you by internet," said Helena in a sarcastic tone. "I know that you're going to be very busy with your son and your life; you won't have time to come here. You probably won't even think about us or the studies when you get back there. You're going to immerse yourself again in the world and you'll forget about your evolution. If you go now, who knows if you'll return, and I don't know how many lifetimes you'll have to wait until you get another chance. In fact, I don't know if you'll ever have it. But I know that you'll do what you think is best for you."

Max was stupefied when he heard this. Frightened, he exclaimed, more loudly than he intended, "No, it won't be like that. I'm being sincere, Helena. I really want to keep talking with you about the studies."

"Look, the studies aren't an extra or a separate part of your life. They're going to be your life. They're a huge commitment. If you make this commitment before the Konceptos it's for this Ansibir and all those that follow. That's why you're here now, since you made this promise before, although you don't remember it. It's a path filled with surprises and joy; there's much to discover if we open ourselves to it."

Max thought about this, remembering the disaster that had been his life a few weeks earlier. Then he imagined the possibilities that he would have in order to understand the world and the functioning of the body; everything that he hadn't been able to learn and know in school or in his research. The advance he could have would be incredible.

He replied, "Then, I don't know how I'm going to arrange everything but can I stay to study with you?"

Helena perceived Max's duality: his arrogance that impelled him to know more and to see what benefit he could take from it, even though another part of him wanted to understand and leave behind the repetition of the errors of the past. Her responsibility as Maestra was to know how to lead him so that he himself could recognize it and overcome it. She had the hope that if he desired it, he would to be able to cleanse himself of his duality and fulfill the responsibility that he had pending.

"Fine, Max. I see now that you're sincere and aren't hiding anything, so go, arrange things and resolve everything you have pending. There should be nothing that remains to cause you anxiety."

Max froze at feeling himself exposed. He measured what Helena's words implied. He felt trapped, but he knew that he should continue. Then he thought of all of the details that he needed to finalize and

204

of the lawsuit. He began to feel miserable again. His mind entered into a confusion of images and fears, but he remembered a certain tranquility he had felt these past weeks and he felt better.

He pulled himself together to hear Helena saying, "You know, I haven't told you that there are certain 'labors' that need to be carried out in some places in the Planicio Terráqueo. As it once fell to me to go live in the jungle in Mexico, maybe it'll fall to you to go to a site, or to several to do something. Check that your passport is in order. It's a good bet that someday you'll have to leave France to go live in another spot, in difficult conditions in order to 'clean up' acts of the past that we have been talking about. Are you ready for it, Max?"

'And what else is she going to come up with?' thought Max, shocked at hearing this last part.

"Well, if that's what I have to do, then I'll do it, Helena." He added, "Is there anything else that you want to tell me?"

"Not that I remember, because everything changes constantly. Just like your way of thinking can change as soon as you leave here. The clarity that you have now will become blurry. The Continuo (Continuum) is very strong and it's always enveloping us." Smiling, she added, "And Max, do you like chile?"

Chapter 12
The Meeting

In the past, man had clairvoyance.
What did he do with it?
Take advantage of the rest and confuse himself.
RAVAGES OF THE PAST

These were the first days of spring, and in spite of the cold at ten in the morning, the sun burned their skin in the dry air of the elevations of central Mexico. Helena, Max, and Diego were excited to see the environs of San Miguel de Allende. The bright colors and the aromas of coffee, chile, and copal enveloped them while they marveled at the old colonial buildings that filled the town center, and at the narrow streets that rose among the hills that outlined the city. The delightful color and enchanting fragrance of the purple flowers overflowing the jacaranda trees filled the small town with a special touch. The three of them left their bags in the hotel and went out like ten-year-old children to explore the place.

"Hey, man, I think I'm going to like living here," said Diego to Max when they saw the center of town.

They walked up and down the cobblestone streets, gazing with pleasure at the quantity of stores, restaurants, and galleries that proliferated in the city. Invaded by the enthusiasm of the moment they were living, the two forgot their differences and decided to go look for apartments to rent. Helena continued walking for a while, thinking about the changes that would come. The three had left almost everything in France to come to Mexico; she didn't know how long they would have to remain in this place, or if Max and Diego would be able to endure the change.

Helena knew that although Max and Diego were ignorant of the transcendent reasons of this moment, in reality they were all living a complicated situation. Now they would have to continue with the twelve labors that her Maestra had begun years earlier; the stay in Palenque was one of them, as was her Maestra Marion's stay in the center of Mexico City, and hers in France. Now they would have to fulfill a mission in this place. She had explained to Max and Diego a few weeks ago about the situation in which they found themselves. "When we arrive at a location, we know a part of the mission, but not everything; this will reveal itself with time and with study. Certain reasons remain veiled until achieving the sufficient konsciousness and submission to be able to perceive and comprehend them. This place has a very important significance which we will have to discover and understand."

Eight months had gone by since Max was initiated in the studies. The relationship between him and Diego was still complicated, and on top of that, Max was going through some very difficult emotional and economic circumstances. In his distress he was spewing his anger and arrogance at Diego, who was increasingly more perturbed by his attitude.

Helena analyzed the situation that they were passing through and thought, 'in order to deserve the Konocimiento we have to cleanse ourselves of the desires and incoherencies that prevent it from entering fully into us. Max and Diego are going to have to pass through many experiences and face many tests in order to receive it. Max has been very committed to the mistakes of the distant past; he'll have to find the root of his negative programming because that's what compels him to act against his evolution. If he doesn't know what his actions and intentions have been, how is he going to end that programming? He'll have to make a huge effort to rid himself of his arrogance. Who does he think he is? And Diego, even though he has great sincerity, doesn't have the clarity or the strength necessary to be able to understand what he's facing. He's a spoiled child who wants everyone to like him. The first thing he'll do is run around chasing girls; we'll how easily he gets distracted by them. I hope he doesn't start forming relationships again, since this site awakens many hidden reasons in people.'

She thought about her own reasons and everything she still had to do. 'It's essential that the Konocimiento be delivered to all of humanity in this HTime so that what happened in the distant past and in the Second HTime doesn't occur again. It's important that everyone knows the truth so we stop delivering our energy to reasons that pull us toward another destruction and step backward for humanity.'

She touched the mole on her forehead, reflecting on the intense changes that would arrive in the coming years. 'There will be battles between those who seek comprehension and submission before the Konceptos and those who continue in their games of dominion and imposition. The Konocimiento of this HTime will give humanity the tools to be able to escape from this evil game in which it has been trapped for so long.'

Helena remembered the yearning that her Maestra had had when the group of twenty people went to live in Palenque. Now, Helena was returning to Mexico to continue with what remained incomplete, as her Maestra had indicated to her. She had to find Pedro, who had been a disciple of Marion. He remained in Mexico studying about the past of the site and about what occurred and altered the development of mankind. This was the moment of reunion with him so that they could continue on together. Some months earlier, by means of Kommunication, it was indicated to Helena that San Miguel de Allende would be the place where this meeting would take place.

Diego and Max put their problems aside and were able to focus on what they had to do. They spent the first two days exploring the town in search of a place to live. They were fascinated with the charms of the city and didn't notice the murky reasons of the location. Diego was amazed by the architecture and the quantity of good-looking young girls that passed him in the street. Although in France he had sometimes studied with Helena in Spanish, he was happy to be able to speak his native tongue with the people, even though at times they didn't understand clearly him because of his strong Castilian accent. Best of all was the sensation of freedom and the desire to relax which he hadn't felt in Remoulins.

In spite of the fact that Max had lost almost everything that he once considered to be his life, he was reluctant to make this change, as he hadn't wanted to leave his son behind. Helena had told him, "It's better that you fulfill what you need to do, and that way you can help your son in a true way. It's the only thing that will benefit him in Real terms." Although it pained Max, he realized that Helena was right.

Helena met with Max and Diego at eleven in the morning in front of the Angela Peralta Theater. The two were excited because after much searching, they had found the "perfect" spot for all of them. They led Helena to a nineteenth-century building situated on a corner where it imposed its presence. Several years earlier the building had been modernized and remodeled into eight apartments. The immense two-story colonial house was painted a shade of sienna and the windows of the façade and the enormous front door were framed in carved *cantera* stone. The second floor windows had small balconies that jutted out over the street. To one side of the front door there was an oval tile typical of the zone that read, "The House of the Coyota." They knocked on the door and heard a sound that indicated that someone opened the electronic lock from above. They heard a woman yell in a hoarse voice,

"Come in, I'll be right down…"

They walked into an ample central patio filled with plants in clay and ceramic pots. An old orange tree dominated the center of the space. There were all kinds of birds in cages surrounding the arches that outlined the patio. Helena was able to see various parakeets, two canaries, and a blue macaw, who in that instant was shouting for his owner to come and see who had invaded his territory. The patio was uncovered and the sky was visible, filled with small clouds that constantly changed form in the wind.

"We thought that this was the perfect place for you; it's full of the plants and birds that you like so much," Diego commented, waiting to see her reaction. Max looked up at the sky, thinking, 'enough with this nice boy act.'

Helena observed the space; it was old and a little worse for wear, but it had a certain charm in the details. The high, thick walls were painted in a pale tone of terracotta that contrasted with the old blue and white tile floors. Helena perceived that the place was saturated with experiences; as it was such an old structure, the walls retained the projections of the people who had lived there before and she felt the energetic load of the astral larvas. She would have to do a good *despojo* (energetic clearing) of these reasons if it were decided that she live there. She felt a warning in her body that indicated to her that the landlady had arrived. She turned to see a sixtyish woman, squat and fat, with her abundant grey hair pulled back in a bun on top of her head. Her face was round and her elongated eyes were almost closed, hidden beneath the folds of the flaccid skin of her eyelids, giving her the appearance of a toad. Her feet were wide, and since she was wearing sandals, Helena could observe that her toenails were long and curved; they resembled claws. One brown eye gazed slightly sideways while the other looked directly at Helena, measuring the quality of these strangers who had entered into her small world. She introduced herself, saying in a haughty and raspy voice,

"Good afternoon. I'm señora Xóchitl, at your service."

Helena thought, 'wow! My third day and I've already run into a demon.' She greeted her kindly, saying in perfect Spanish,

"Good afternoon, señora. We came to see if you had some apartments for rent. My friends told me that they were very clean and well-equipped. Could you show them to me, if you'd be so kind?"

Xóchitl looked at her, trying to decipher what it was that seemed so strange and unnerving about this woman. She didn't seem Mexican, but she spoke Spanish as if she were. She possessed a calm and security that bothered her. She tried to size her up, noting that she was well-dressed but in a simple manner that didn't call attention to her. But when she noticed the sparkle of light which leapt from the emerald ring that wore on her hand, she thought, 'ah, at least this one will be able to pay the rent.' She, obviously, was the leader of the trio. She kept looking at her and her distrust returned; who was she and why was she running around with a young Spaniard and a handsome Frenchman? This didn't seem at all right to her; who knew what they'd be doing together? But she needed the rent money and relented, even though the thought that she was going to regret having rented rooms to this group passed rapidly through her mind. She took her keys out of the pocket of her sweater and said in a high-handed tone,

"Follow me, please."

They walked up the stairs to the second floor. When they entered to see the apartment, Helena asked and the Konceptos indicated to her that she should rent it; it would serve as a home but also as an office. The space was small, but it had good light and she could make use of it to attend to the people who would come for healing. Before leaving France she had understood that she should dedicate herself to healing and to teaching, in addition to finishing the book she was writing.

The furniture was old and worn, but she decided that she could revitalize the atmosphere with a good cleaning, new cushions, and a little paint. Max and Diego understood when they asked that they should share the other apartment which had two bedrooms. They didn't have much money and even though they really didn't want to live together, they accepted. It was very practical, and besides, it would be easy to study there with Helena. They sealed the deal with Xóchitl, who accepted in spite of her misgivings, hoping that they wouldn't cause her any problems. To calm herself, she told herself that San Miguel was full of odd people, although these had something different that disquieted her.

That afternoon, Helena, Diego, and Max moved their belongings from the hotel to the apartments, installing themselves in that peculiar place. The days went by and they adjusted to their new environment. Xóchitl was still wary; they caught her all the time as she passed in front of their doors, trying to discover them in some clandestine activity.

Helena fixed up her apartment in a simple manner, painting certain walls of the house in happy, bright colors, the ones she had always liked in Mexico. She bought various cushions and decorations in the artisans' market to liven up the space. She found some embroidered curtains and hung them in the bedroom, tossing out the old ones which were worn out after years of use. In order to give a final touch to her decoration, she filled the house with plants.

Two weeks after their arrival in San Miguel, Helena understood that she should go to the school of Bellas Artes that afternoon. She walked to the old school and entered the huge central patio, which was full of artwork and vegetation. She sat at one of the tables that surrounded the garden with its central fountain. She ordered an espresso and took out her cigarettes and the notebook she always carried in her bag. The breeze rustled the leaves of the giant bamboo which dominated one corner of the garden, producing a harmonious sound. She opened her notebook and began to study.

While Helena was absorbed in her studies, a man watched her. He noted the manner in which that enigmatic woman distractedly touched the mole in the center of her forehead. She was focused on writing something in her notebook. He perceived an unusual energy in her and it caught his attention immediately. He decided to investigate to see if his conjectures were correct.

Helena, absorbed in writing, noticed a shadow fall over the table, distracting her from her thoughts. She lifted her head to see where it came from and saw a tall man with a felt hat standing in front of the table. He had a wide grin that peeped out from behind his beard. His brown eyes twinkled like those of a naughty schoolboy.

"Hello, you must be Helena, if I'm not mistaken. I'm Ped-ro Castro Cárdenas, and I believe we have something in common." He extended his hand to greet her with an air of studied elegance.

"Ped-ro?" Helena understood instantly who he was and smiled, inviting him to sit with her. He was the man who her Maestra had told her about several years earlier. She couldn't get over how he had pronounced his name with a certain pride and the strange way in which he had dragged out the syllables, making it seem like two names. "But how did you recognize me?"

"Well, I don't know many people who have that singular mark," he answered at the same time that he pointed to her forehead. He smiled and sat in the chair that Helena offered him. "In fact, when I was studying a few weeks ago, I understood that you were on your way here."

Pedro was approximately fifty years old, although he seemed younger. He took a silver cigarette lighter from his pocket to light his mahogany pipe, from which the sweet aroma of fine tobacco disseminated. His long, dark brown hair was tied back, making his smooth, pale skin stand out. His eyes demonstrated a mental agility and an eagerness to absorb and analyze everything around him. The combination of his intense gaze and the angle of his narrow nose gave him the appearance of a hawk. He was dressed in a white shirt, jeans, and a designer-label navy blue sweater which he wore tied around his neck. His attire resembled that of a professor, and in fact, although he considered himself an author, he had made his living teaching contemporary literature in a private school in the capital. When he spoke, he loved to reference Latin American literature. He had arrived six months earlier in San Miguel with a group of people that he gathered together in the years since Marion had gone.

Pedro was very curious to know the stranger who had recently arrived. Marion had told him little about her, but from her remarks it was obvious that she had great confidence in Helena. When Marion departed, she told Pedro that one day Helena would return to Mexico and in that moment they would have to join forces in order to fulfill a very important reason. Pedro was sure that Helena was going to help him advance.

Helena replied, "Yes, a few months ago the Konceptos indicated to me that it was time to return to Mexico. I understood that this would be the meeting place for everyone related to the studies. But tell me, Pedro, what have you observed of this place?" Helena and Pedro spoke a long while about the subject. Pedro offered his help in getting fully adjusted to the city.

"I imagine that it's been a big change for you after living in France. If I can help you with anything you need, don't hesitate to contact me. I know a lot of people here. I work in a bookstore, and it's been the perfect spot to meet people. In fact, there are folks who come from all over to talk with me." He said it with an air of arrogance that took Helena aback. "Maybe we can meet, you and I, to talk about what you've studied, and we can share the information that we have. When can we see each other?"

Helena responded cordially but thought, 'he speaks of the studies as if it was a social club, and I think he saw me with certain envy. I hope that I'm just imagining these things, because we came here for a transcendent reason.' Her memories flew back to the time she had spent in Palenque and to how things had unfolded there. She sighed, wishing that this time they would be successful.

For Max and Diego the first months in San Miguel had gone by in very different ways for each of them. Diego found, almost immediately, work as a waiter in one of the many restaurants in the center of town. For Diego, the kind of work he did hadn't much importance; he was accustomed to doing what was necessary to get ahead. In time, he met many people through his job, among them, various young women who were attracted by his energy.

When Helena saw that he was very excited about the situation, she told him, "Ah, how easy it is to become confused in the world; a little energy and we think we're pretty impressive. The energy that each man is given should be used for his evolution." She explained to him that the interest that the girls showed in him wasn't because he was the 'max,' but because of the presence of the Lux that he received from studying. People perceived it, and their attraction was to it, as she had explained to him months ago in Remoulins. Nevertheless, Diego hadn't wanted to accept it and was carrying on distractedly, going to bars almost every night and taking importance away from analyzing the place. Helena warned him that if he let himself be carried away by emotions and sensations, he would attract reasons from a lower Astral and this would end up affecting him.

San Miguel was a place full of strange occurrences and eccentric people. In the four months since their arrival, Helena had gained a certain recognition in the community for her ability to heal with emanation, and for the things that she talked about. She had had contact with individuals who belonged to the diverse groups who lived there, and upon speaking with them she had noticed various reasons that were then clarified for her through study. Helena understood that many people who came from other places were attracted by the energy of the site. In several cases they were people who fled from a life that didn't turn out like they wanted. Without realizing it, they returned to this place which in the past had an importance for them. Others had come because they felt an affinity with the energy that had been expressed there.

In San Miguel, as in many other locations that had a previous development, the experiences and thoughts of the past had left an energetic residue that still resonated in the present. Expressed in this residue were the intentions of the earlier races and civilizations that had lived there. These energies affected the people who arrived in the present, but in different ways and according to their affinities with these old reasons. In many cases, when they installed themselves in the spot, something curious happened to them. They "reinvented" themselves, projecting the desires and interests they carried inside and that were part of the habits and programmings of their previous lifetimes. They acted like children playing dress-up in the old clothes they had found in the attic, disguising themselves as artists, investors, or shamans. Thus, reinvented and free to do what they liked, they gathered together in groups according to their affinities, and in general, the energies drove them to repeat their earlier programmings. And even though the deeds weren't repeated exactly, the intention was the same: re-create an illusory world in which the Ego distracted itself in its own fantasies.

Helena had understood that San Miguel was not simply a meeting point for many Egos, but that the energies there acted in order to activate the negative programmings of each person. This would be the battle that each one of the initiates in the studies was facing. They needed to understand how the site was affecting them in order to be able to reject their programming; if not, each one of them would once again repeat the same errors as always. She saw that Max and Diego were very vulnerable because each had a great emotional imbalance.

One day she explained to them that they needed to be careful, since in moments of great unbalance, they could open themselves and absorb the negative energies of the area or deliver their own energy to fuel the reasons and the Egos that opposed what they needed to carry out in the place. So that they would understand more about the subject, she read them this study: *"**Human beings perceive an environment, sometimes they are aware of it, but normally they lose themselves in it, entering into a false ensoñación*** (*state between dreaming and waking) *that allows that their energy be used. [...]The Ego is an apparent emptiness which is filled with the experiences of the environment"* (*17 Mar. 90*).

"But how is it that the people who live here and the environment itself are opposed to us?" Diego asked her.

"Remember that Egos have no nationality; they arrive at a location because they feel or remember something, even though they aren't konscious of it. The Egos that are born in an area, and who are attuned to these reasons of the past, are the 'guardians' of the place. They're there to guard and defend a site, for example, a 'place of power' or a *boca* (outlet) of energy, so that the energy which arrives there is only used to keep on recreating their fantasies, which generally are just repetitions of what happened in the past. To achieve this, they will attack and will do everything possible to stop that which represents a change in their folly."

216

Incredulous, Diego asked her, "But how, if they're really good people? What do you mean, Helena?"

"You're just chasing after skirts and don't realize what's going on."

Max exploded in laughter, enjoying the moment, but the truth was that he too was astonished to see how much Diego had changed. He walked around with his head in the clouds, hooking up with women that he met while working in the restaurant, and Max knew that he was using drugs. Diego was so agitated that it was impossible to hold a conversation with him, much less talk about the studies.

"When I say that they attack us, I mean that their energy is contrary to ours. It's a constant energetic battle because they don't want a positive reason to overcome the erroneous one," she replied. Helena perceived Max's thoughts. She thought how easy it is to see other people's faults, but seeing our own is more difficult. She knew that when a *pequeño takes drugs he loses and weakens his Lux, since he leaves open channels through which negative entities can make use of his Lux. In addition, he disrupts the other pequeños (2 Feb. 87)*. She had spoken with Diego, but this was something that he should have analyzed and understood by himself. Diego hadn't wanted to accept how energy functions between people and in the environment, or how his actions could affect the group. If things kept on like this, they were going to end up resolving themselves in a drastic way. Each person is placed in a situation so that he learns, but when it's not done in its indicated time, it becomes more complicated until it reaches a drastic resolution which shakes up the Ego so that it reacts.

For Max the transition between France and San Miguel had been very difficult. On top of the emotional upsets and all of the changes in his life, he had to find work to pay his expenses, and although he really had no other experience, he understood that it wasn't beneficial for him to continue on within the reasons of medicine. Since he couldn't speak Spanish well, he ended up teaching French in the school of Bellas Artes. But this money didn't stretch far enough to allow him to dine out in restaurants as he would have liked; this small pleasure was one of the things that he most missed about his life in France. In the beginning, Diego, trying to make him feel better, had brought him leftovers from the restaurant, but this gesture only humiliated Max more. His situation had provoked a great emotional instability in him. He was accustomed to people recognizing his intelligence, but now he was living in circumstances where he had no position that gave him importance or power. He missed his son, who no longer wanted to talk to him; Michel had just said that he was crazy for having abandoned him and everything that he had to go live in the middle of nowhere. Little by little Max's anger and resentment came to the surface and he began to act in a domineering way toward Diego and Helena. His attitudes demonstrated an air of superiority and worthiness, something that he had done throughout countless centuries. This produced many conflicts within the group.

One day Helena took him to a park so they could talk calmly. They sat in front of a basketball court, listening to the cries and laughter of a group of children who were playing. She explained to Max that everything he was living were his tests; in his case it was necessary that he pass through a drastic change so that he could have an adjustment in the interests that he had had for so long. Max was very resentful that day; he hadn't wanted to listen to her and began to complain.

"But you have to understand me, Helena. I left everything behind. It's not like in Diego's case, where he only had a little shop and didn't even want to be there; and here he's having a great time. I had an important career and a son who needs me. I had to leave an entire life. And now what do I do? Teach French to some spoiled brats. I know, the changes were necessary, but it doesn't seem fair to me that I had to give more than Diego to be in the studies."

Helena wanted to halt Max's fit of pique; he was only going to drown in his emotions and with that he was going to lose the little clarity he had. She said, "You're old enough to be able to take responsibility for your own actions. I want this to be absolutely clear to you, Max. It's not that one of you left more than the other or that one gave more. Each person should give what corresponds to him. For you, it may seem that Diego gave very little, but for him, it was his life. It hurt him as much as it hurt you to leave yours. We don't judge ourselves with human eyes, Max. Diego may be thinking the same about what you gave." Max crossed his arms, offended by Helena's lack of sensitivity. "Look, what you're living right now will pass as soon as you want it to pass, and then another adjustment will come and then another. Only if we accept what needs to be lived can we pass on to another lesson. So if we oppose the changes, they become more chaotic. You'll have to watch your emotions, since you know what they involve."

That night she gave him a study to help him understand. *'The reason for negativity is fear: man does not find the path. He knows the truth but he has to make all of the multiplicity of konsciousnesses within him fit together in order to understand it. He has to enter into a spectrum of negative reasons as from dark to light, pass through mistakes or sufferings which are deficiencies in order to comprehend. When one understands since the stone a selected one is formed, but he still has a remnant to refine him even more.*

Upon leaving the Planicio Terráqueo the Ego will have a clearer appreciation of its experiences, it will not think of the hardships it had, so it will understand. The same occurs in the Trebolo, an Ego in completeness forgets its previous bonds and stops having the

perception of the emotions that it had in the Ansibir, it observes them without getting excited. In the Ansibir it happens in the same way, what tormented the child loses importance as an adult. Thus we have to remember what truly is important' (26 Aug. 89).

Even so, Max was unable to find a balance; he fluctuated between explosions of rage, resentment, being fed up, and other feelings. In the moments when he was very upset, Helena advised him to do physical tasks, above all cleaning, so she sent him to clean up the apartments, saying that it was wonderful for decreasing the arrogance of the Ego. Despite Max's efforts to control himself, things got worse until one day when he was in the kitchen of his apartment; Max became angry over an innocuous remark of Diego's. The glass door of the microwave exploded just as Diego opened it. He wasn't hurt, but they both suffered a huge fright when they understood that it was Max's rage that provoked the "accident."

Diego yelled at him, invaded by the feeling of fury and fright he felt because of what happened. "Damn it! Man, look what you did! I'm fed up with your tantrums. I don't know why you treat me like an idiot all the time. If you don't want to be here, why don't you go back to France and you can keep on feeling like the eminent scientist Dr. Know-it-all who was fired from his job for being an asshole."

"And what about you? You think you're so hot because you're hooking up with half of the women in San Miguel? And now you're in love with that Laurita, who's only making a fool out of you; you'll see, one of these days she's going to give you a little surprise…"

Diego left, enraged. A few months earlier he had gotten it into his head to look for love and had recently begun a relationship with Laura, a 21 year-old girl born and raised in the town. Helena had observed that she was a huge distraction for Diego, since she was demanding much of his time and energy. She had no interest in what Diego was studying; she dreamed of having a traditional life, and she was pressuring Diego to live with her. Helena and Max had noticed the doubts Diego was feeling and how, little by little, he was becoming more involved in the relationship, even though he thought he was managing things well and could control the situation.

The day after the microwave incident and discussion with Diego, Max talked to Helena; she clarified for him that he projected his rages with great force, and that his negative thoughts were affecting them all.

"How is it possible that I made that thing explode? I know I was really angry with Diego, but not enough for that. However, I understand that I was the one who provoked it."

"We know very little about how energy works, Max. Emotions are like a detonator. You have a lot of stored-up anger and it all stems from your arrogance. For a long time you've thought that you're the greatest and you've wanted to have an importance before the world. You're afraid of losing it, of being a nobody, without position, without money, without the praise of the public. You don't know who you are as an Ego, you only see the image that you've forged and you're afraid of it dissolving because you haven't as yet projected a new, positive foundation. It's in these moments of imbalance that you open yourself, and you become a duct through which contrary energies can manifest themselves. It's precisely then when those energies possess you. But with study you can adjust everything that is troubling you and avoid possession."

"I'm possessed? *Ça va*! Now you really are mocking me. Please, Helena, I don't believe in those things. And it surprises me to hear a highly educated woman like you talk about them."

"We're possessed when we allow our own confusions and dualities to dominate us. We know that a contrary thought attracts a similar one and this attracts another until it forms a chain which we can no longer stop and it ends up projecting itself in a deed. An Ego opens itself in a moment of great confusion or distress and receives or takes from the environment the energies that are attuned to that moment of negation. And then it projects them with great force in the Astral level, causing an event in the same Astral plane or in the physical plane, as happened with the oven.

Look, I'll read you this study. *'A person in disharmony can disturb another who is not in his exact balance, since he "expels" that energy*

in disharmony. That world which is matter always expels, it is an escape of photons, within a normal range, but a person in imbalance expels "projectiles" that alter the rest' (1 Apr. 89). Everything that we think, feel, say, or do is sound. This place is saturated in the Astral dimension with the thoughts and acts of the past. *All that surrounds us is projected in us and we in what surrounds us. All presence saturates with sounds, larvas, etcetera, all action good or bad, all sound or movement forms larvas, whether they be convergent or divergent (24 Mar. 90).* That is, attuned or not to our reasons."

When they were all together that night in class, Helena told them, "Each man is born with two programmings: one positive, that is, that which he has to fulfill in the Ansibir in order to continue in his evolution, but also a negative one, which is that which holds the harmful trajectory of the centuries, what the Ego has repeated over and over and which has distanced him from his comprehension. If not, we wouldn't be here."

"Then the negative is already within us, Helena?" Diego asked, uneasy with the discussion.

221

"Look, Diego, negativity is that part of each one that doesn't want to fulfill our reason for existing. The negative is that which doesn't arrive at the purpose for which it was projected. In the Ansibir we all receive opportunities to correct this fault or deficiency in comprehension within ourselves, whether it be in large or small details. *The circumstances through which an Ego passes are related to this deficiency; each one is pressured where he is most lacking, in what affects us, distresses us or matters too much to us, for an apprenticeship, in order to resolve this deficiency. When this deficiency is abolished, the circumstances change, the Ego has understood (28 Jan. 89).* But our path has been complicated by the commitments that we have made with other Egos and with the whole of humanity. You, for example, have understood that your problem is the bonds that you create with other people and that these ties have detained you. But even so, you insist on making bonds with the people you meet. Now you have to examine the reasons that induce you to repeat the same. You have to examine your relationship with Laurita."

Diego became alarmed at hearing this. "But, Helena, I love her. I've even thought about marrying her."

Helena stopped him and said in a harsh tone of voice, "You know that you can't marry someone who isn't in the studies, right?"

"You've told me that, Helena, but I don't understand why," he answered, on the defensive.

"You should have understood by now how energy functions between Egos. Look, since you are studying, you receive a greater charge of energy. You have an elevation that you should use to pull her upwards. And here it's happening the other way round: she, at not elevating herself, is pulling you downwards. That's how energy functions; it has nothing to do with feelings."

"Well, I don't know what I'm going to do, because I want to be with her," he insisted.

Max looked up at the ceiling, not believing what he was hearing. Diego really had lost his way.

"There are only two things you can do: either she decides to study because what matters to her above all else is her evolution and not her relationship with you; or you separate yourself from your evolution and submerge yourself in karmic reasons with her. You decide, Diego. In one hand you have the Konocimiento and the most transcendent that there is, and in the other you have a transitory comfort that will last a short while." She paused and added, "You've been in this position in the past, don't you remember? *For that reason it is important to have the konsciousness of what was in the past, the sequence previous to this Ansibir so that we understand our current characteristics; why we invert the powers that are given to us, why are we our own worst enemies, why we direct ourselves toward a reason and then conspire against that same reason. Part of the rebellion is a self-destruction: a part of the Ego tends toward survival and the other to destroying itself, and this is reflected even in the most trivial acts of our life. We love ourselves and hate ourselves at the same time [...].The older the Ego, the harder the road is because of the additions that it has from the previous centuries, which it has not been able to overcome"* (17 Sep. 88).

Diego didn't answer; he was worried and irate because he needed to make a decision.

Helena directed herself to both of them. "It's very important that you understand something. There are very intense battles coming between the Egos who seek their evolution and those who don't want men to receive the Konocimiento or that they understand the truth. It's in the moments when we're distracted, weak, or divided amongst ourselves when they are going to attack us. It's essential that you both understand how negativity acts in each of you so that you can reject it, and thus not be affected by the reasons of the past or by your fears." She paused, thinking about the events in Palenque and continued. "I've seen before what can happen when people let themselves sink into their negative programming and I don't want to live it again." The image of Palenque and then the face of Pedro passed through her mind and she said to herself, 'I hope that the past doesn't repeat itself...'

Chapter 13
Chamucos and Chichimecas

There was an arrogant Prince, who feeling himself powerful,
gathered his supreme Huestes, sought to take over the realms that
circumscribed, but he lost the battle and was
locked in a sphere with all of his Huestes;
they haven't understood their defeat, or the lesson that this brings
and he wants to begin a new battle;
don't help him, please, cretinism is contagious.

MEDICAL RESULTS

Everything seemed to be accelerating toward a point which Helena was still unable to distinguish clearly. The instability of Max and Diego worried her, and moreover, the distance and lack of interest that Pedro was showing in the studies. As if this were not enough, San Miguel seemed to have become agitated. It was not only the discordant energies of the place that were affecting the study group, but also that each one of the initiates had attuned themselves to them. With this, the contrary forces began to manifest themselves. Strange things were happening: a window in Diego's bedroom broke mysteriously, some men assaulted Max in the street, and a few nights earlier, while the group was studying in Helena's apartment they heard the murmuring of various persons in the hallway. Suddenly they noticed the sound of a door slamming loudly and laughter resonating in the stairway. They heard such a commotion that they swore there was a group of intruders outside the apartment, but when they went out to check, they didn't see anyone. Almost immediately the electricity went off in the apartment, leaving them in the dark for a couple of hours.

"Honestly! How they pester us with their attacks, they just want to annoy us," Helena remarked to them.

Diego and Max looked at each other, dismayed; they hadn't expected this facet of living in San Miguel. Max continued on without wanting to accept how things could manifest themselves in other levels which are beyond the physical.

A few days later, Helena went out in the morning to go to the open-air market. The sun was very intense and she was grateful for the shade underneath the tarps that covered the stalls. The market was full of people who wandered through the narrow aisles, looking at the tables packed with all kinds of articles. She stopped at a stand filled with fruit: juicy strawberries, mangoes, papayas and other delights. She was chatting peacefully with the vendor when she was approached by an old indigenous woman who was begging for money among the crowd. Ugly and stooped, the woman walked slowly towards her with a wooden cane in her left hand. She was dressed in filthy rags and her head was covered with a shawl. When she was almost in front of Helena, the woman looked up. What Helena saw in her face left her dismayed. Her eyes were brown, but there was no lux in them. Her dark face, wrinkled and twisted, showed great anger. She extended her bony hand towards Helena, demanding that she give her money. Helena told her gently that she couldn't give her anything. She knew that this woman's condition was due to her karma and if she helped her by giving her something then she would take on part of it as well. The woman insisted and insisted, moving closer and trying to touch her back. Helena turned to tell her to leave and stop bothering her; she wasn't going to give her anything. The woman raised her cane and pointed it at Helena's face, cursing her under her breath, and demanded, "Give me some of what you all have. Where are the others that are with you?"

Fed up with the woman's insistence, Helena exclaimed, "Get away from me, demon, go and leave me in peace." Helena knew the woman was trying to reach her because of her Lux and that was why she wanted to touch her.

Suddenly the face of the woman was transformed. Her eyes shone like those of a cat in the night, and she pulled herself up erect and strong. Full of fury, she raised her staff and demanded once again of Helena in a deep voice, "Give me some of what you have." The image of a defenseless woman evaporated and a presence filled with fury and aggression took its place.

With an even firmer tone of voice Helena said, "Get out of here! If you're possessed, it's because you allowed it; go away and break with it." Helena began to emanate energy, allowing it to flow through her hand in order to stop the attack.

When the old woman saw that she could do nothing, she hunched over again. It seemed to Helena that upon feeling the emanation, the woman's body had diminished and the presence that had possessed her abandoned her. The woman withdrew, writhing as she walked.

The fruit vendor was stunned. He couldn't believe what he had seen and how that woman had changed in appearance. He knew her and he told Helena in a whisper, "She's not a beggar; she has family and lives a few blocks from here. Look, it's best not to get involved with her. She's known to be a malicious woman and some people say she does witchcraft."

'Please, woman?' thought Helena. 'She's a demon and the whole world knows it.'

This encounter was a warning for Helena that something powerful was happening in the town. It was incredible how these presences felt her movements and perceived what she did. They even knew of the group. The same thing had occurred years ago when she was with Marion. The demons possessed the people attuned to their reasons and they arrived seeking the Lux which they lacked. Not having it made them angry but at the same time they attacked it because perceiving it was painful, since it revealed what they really were.

Helena arrived at the apartment and called Diego, Max, and the rest of the group to tell them what had happened. She thought, 'we'll see if with this, they realize that with their own actions they're helping to activate these dark reasons.'

In the evening after they got off work, they all gathered to study about it. The only one who didn't come was Pedro. When they were all arranged in their seats, Helena commented to them, "What I witnessed today with that old woman was a possession by a demon of great strength."

When they heard this they all looked at each other, confused. One man in the group questioned her in a skeptical tone of voice, "But Helena, demons don't exist, it's just a fable. This type of possession only happens in the movies, not in real life. Or do they?" He began to doubt his certainty at seeing the seriousness of Helena and the others.

"Yes they do; they're Egos in rebellion that have rejected having bodily matter. They seek to lodge themselves in the matter of an Ego, whether it be human or animal, which lends itself to this. Look, people of this time don't speak anymore about these reasons like they did in the past. If we could ask the Spaniards about the things they saw when they arrived at these lands, we would be amazed by the practices that the native peoples had. Since prehistory battles between angels and demons are spoken of; the church continues performing exorcisms, and well, in these parts it's their daily bread and butter. It's something very old, that an Ego allows itself be possessed by another Ego without matter or by an energy in the environment, to see or do something that it normally can't, as in the example of shamans and sorcerers. In the same way, an Ego without matter needs to take the energy of an Ego in matter to make itself visible and act in this dimension. The ceremonies carried out by many indigenous cultures, for instance, serve to call the demons, that is, their hierarchs or leaders that they adored before. They attract them so that these enter into a matter which has an affinity with them."

Someone in the group commented, "Shamans take hallucinogens to enter into trance and then talk to spirits. Is that the same thing?"

"Yes, look, there are various types of possession. The one we talked about a few days ago is possession by a person's own reasons. These are what we should be concerned with studying. But there are other kinds of possessions."

"And are there things other than drugs that can open us to these reasons?" asked Max.

"Yes," replied Helena, turning to look at Diego, who ducked his head, pretending to be writing something in his notebook. Helena continued, "Any reasons that distress or shake us up can open the sieve of our matter and allow another energy that shouldn't be to enter; for example, when we get drunk, or when we have a very strong emotion, or upon having sexual relations with a matter of the same gender. And it is also in that moment when some diseases can enter.

Generally, it only happens when an Ego in matter yields his to an Ego without matter, whether it be an Ego Trebolario or a demon (an Ego that escaped from the stone before its time). Most of the time it's because of pacts made in the past. These possessions occur when the matter is disturbed and detached from the Ego, in moments of weakness provoked by intoxications, trances, or grave or emotional illnesses, for example. In the case of mediums we see how an Ego permits an Ego Trebolario to enter into its matter when it's in a trance. *A medium (a half unit) provokes a decay in himself because the vital fluid is being absorbed by the Egos Trebolarios that he lets in. He suffers an organic and psychic imbalance; that half does not evolve and produces a great disharmony because of all that it attracts (29 Mar. 88).* The motive of lending matter to another Ego is a crime and brings as a consequence a great karma. These possessions provoke a very grave damage to the matter of the person who lends it; they can remove its vital energy, or make it become ill, or affect it mentally.

228   There are other cases of people who open themselves consciously to channel supposed 'guides' or 'spiritual masters'. But we know that a master would not enter into matter that way, since he would know very well that this act isn't permitted because it goes against the personal evolution of the Ego, and is a crime. Imagine the things or ideas that those Egos Trebolarios or the Desideratos could introduce into the thoughts of the people they enter into."

Another person in the group asked her, "Helena, I've heard that ghosts, well, I mean to say Egos Trebolarios, many times ask people for water and complain that they're thirsty. Why is that?"

"It's not that they want water or that they're thirsty, since they no longer have matter, but in reality what they're asking for is the Lux which was taken away from them due to their wrongful actions. The Lux is represented by water, which is an Eternal element. Ok, let's return to the theme of possessions."

"What is a desi…deshidra…?" stuttered Max.

"Desiderato. It's what people call a demon. They are Egos that have rejected having matter and the apprenticeship in the Planicio

Terráqueo. They are those who left, or were released from, the stone before they completed their lesson in it. That's why they continue in rebellion. They lodge themselves momentarily in the matter of someone who allows it, provoking a great damage to them in the process.

A demon is an Ego that in the reasons of perturbation or confiscation, remembers the powers of the past, and attracts for this reason its own causes and diverse attitudes. The demon should be called Desiderato, who are the leaders of the past in the Kosmo of the Trinak, and the demons can be us when we are in the error. *They do not have matter, and due to their own reasons, take one that is favorable for them or that accepts them, desiring in this, to resolve its own misfortune, this is, its perturbation. The favorable matter is that in which there is a marked separation between itself and the Ego that inhabits it. The Ego separates itself from the matter in the face of its helplessness in events or because of intolerable physical pains. It is for that reason that in this matter is registered the forceful entry of the Ego desiderato, that is, the possession.*

A desiderato may give a disease to the matter since it affects the nervous system and reduces the pulsations of the Ego, provoking painful states in the matter. There are certain desideratos which have the propensity to take possession of animals. This can be when the animal (the Ego that protects it) varies according to the governings of the moon" (5 Apr. 91).

"And who are these Egos, Helena?" someone asked.

"They're the Egos who in the past were the most prepared and who had a higher hierarchy. They are the leaders of the Huestes (Hosts)."

"Can they introduce themselves into any matter?" someone else asked, worried by what he was hearing. He thought, 'oh my, what's going on here?'

Helena grasped the uneasiness of the majority and clarified for them, "In order for this to happen there must be an acceptance on the part of the Ego who lends its matter. If not, the possession isn't possible. And it is *only if it lacks matter (the one that wants to introduce itself). An Ego in the Ansibir cannot abandon its matter and occupy another.* [...]*There are cases in which an Ego in the Ansibir by previous agreement and by means of a complex process*

occupies the matter of an Ego that is on the point of Trebolo who cedes it to him (19 Nov. 88). This is what has happened in some cases called 'miraculous' when survivors were found after a catastrophe in which they have been buried up to fifteen days without food or water."

A few days after this class, Diego returned to the apartment at dawn, dizzy and worn out after having spent the night drinking. That night he had worked a double shift and then gone out partying with Laurita. When he reached his bed, he fell into it, exhausted. It seemed to him that he had barely closed his eyes when something woke him.

Diego felt that his bed had moved; he had the sensation that someone had sat upon it. He didn't know if he was asleep or hallucinating. He tried to open his eyes but couldn't; fear ran through his body. Very gingerly, he extended his right arm to see if there really was someone or something on the bed. To his horror, he touched something solid. He turned his head, terrified and ready to run; he saw that a figure was lying by his side. He perceived that the form wasn't precisely that of a man; it was greyish, without a defined face. He wanted to shout but no sound came out of his mouth. The presence forcefully grabbed his right arm and Diego could no longer move. In that moment the figure held him down by the neck and started to strangle him. Diego tried again to yell but without success. He fought to get it off of him, but his body was paralyzed. With the hope that it was only a dream, he desperately tried to open his eyes. He tried to call Max so that he would come and help him, but he remained mute. His panic grew when he sensed that the presence was removing all of his vital energy. The figure placed all of its weight on top of Diego and continued suffocating him. With a great effort, Diego battled to gather his energy and was able to raise his right hand to emanate a Lux. With this emanation, the presence disappeared and he awoke, bathed in sweat. He struggled to breathe. He tried to turn on the lamp by his bed, but he couldn't control his hand and it fell to the floor.

Max heard the noise of the lamp as it fell and went to see what had happened. Outside of the bedroom he called to Diego, but he didn't respond. He opened the door and turned on the ceiling lamp. He saw a scene that left him cold: the lamp was shattered on the floor and the bed in a shambles as if he had fought with an intruder. Diego lay languid on the bed, pale, and soaked in sweat. When Max ran to check him, he saw that Diego was half unconscious and had a long, deep cut on his face. While Max tried to stop the bleeding, he tried to revive him, shouting, "Diego, Diego…what happened? Who attacked you?"

Diego continued half unconscious and without the energy to speak.

The birds below in the patio squawked and the dogs in the street barked at the commotion they felt come from above. Max went out to look for Helena, who was already walking down the hall. She arrived to find Diego prostrate in bed and without vitality. She performed a *despojo* removing the energy residues that the desiderato had left and she emanated energy with her hand to give him strength and heal him.

After a period which seemed eternal to Max, Diego recovered enough strength to be able to tell them what had happened.

"The emanation saved your life. Good, Diego," Helena commented to him.

Diego was still very weak, and Helena explained to him that the desideratos eat the vital energy of the Ego, but with the healing she had performed, little by little he was going to recover. Max went to his room to grab his emergency medical bag.

"Leave it, Max, the emanation is going to be sufficient."

"But Helena, can't you see…?"

"When are you going to realize the magnificence of what you're learning, Max?"

"What's going on, Helena? That thing almost killed him!" Max demanded with the anger born of his fear.

"Come on, Max, one scare and you're ready to head back to Paris?" Helena remarked while Max looked at her, confused by her attitude, but her words were stronger than the fear he felt. The words that came out of Helena's mouth, full of assurance, helped Max get rid of this unwanted feeling and he could think clearly.

Once Max was able to calm down, he asked, "Why did it attack him?"

"Diego exposed himself when he broke the protection that he has. He didn't take seriously what was said to him about how his actions were leaving him vulnerable. He kept taking drugs, and maintaining the relationship with Laurita, which instead of helping him, left him drained and exposed. He opened himself to these energies when he lost himself in the environment and distanced himself from the Voluntad."

In that moment they heard footsteps in the hallway outside the apartment. Max went to open the door and peered out; he saw their landlady Xóchitl slinking down the stairs blanketed in the darkness of the night. Xóchitl feeling Max's presence, turned to look at him and her brown eyes glittered in the reflection of the moonlight that entered the central patio. She smiled and Max was able to hear her mutter something; even though he made an effort, he couldn't make out what she said, although he swore that he heard her laugh while descending stealthily to her room.

Max returned to Diego's bedroom and told Helena what he had seen. She explained more to them about the battle in which they were involved.

"What the demons and their followers want to avoid is that man understands. They're going to use all kinds of ruses to distract us, weaken us, and stop us. And why do they want to stop us? The ones that attack us now are only the *achichincles*. The true intention of the leaders of the past is to immerse everyone in darkness so they can manipulate them and control them once again, as in the past, to a new rebellion. What they want is to escape from matter and from the Planicio Terráqueo, but they don't understand that this won't be allowed, they won't arrive at anything. Nevertheless, we are all connected, and if they gather together enough strength in their deranged plans, they are going to pull everyone towards a disaster. It's like a crazy person who decides to throw himself off a building: if the head, that is, the supreme leader, decides to jump, the legs, the hands, the chest, and the whole body follow him also.

But they should know that *the Reason Life is ruled by imperturbable laws, by reasons that were delivered,* since their formation, and that are beyond their small awareness and their demented plans.

It is thus that the impetuous fight of the leaders of the Huestes and the supreme hierarch *will not proceed in a triumph but in a disastrous defeat in which we will all be involved, the conscious and*

the unconscious, but the stubbornness of the demonical forces will not be stopped unless the little humans, my brothers, stop playing with their little balls and their material dreams and understand, and rebel giving finally their immediate and absolute submission to the Real, it is because of this that each and every one of us must walk on the road of the Konocimiento" (16 Mar. 08).

Chapter 14
From Helena's Book:
Where are we going to?

After the intense events of the last few weeks, Helena was grateful for the chance to have a moment to focus on a more pleasant task. She sat at her desk and reflected on the third question that she wanted to reveal in her book: where are we going to? Thinking it over, she said to herself, 'the future can be beautiful, full of harmony and contentment, if we surmount the lessons that are being presented to us now. If we fulfill all of the steps of our development, at the end of the Seven HTimes we Egos will leave the Planicio Terráqueo and return to the Kosmo of the Trinak to finish what was left incomplete there. Afterwards, we'll pass through new and astonishing stages in our evolution within the Fusion, which for the moment we are incapable of understanding.' She opened her notebook and wrote:

> Our evolution is integrated into the process of development of the Planicio Terráqueo and it into that of the Kosmo of the Omnipotente Jehovah, and the Kosmo in turn into that of the Universe or Fusion. Each one continues on in its own evolution, but there are specific moments in which great changes and transformations are manifested that affect the All. We will see astonishing modifications. The Earth will experience changes in its appearance, new continents will emerge; new races of men, new species of animals and plants, colors in the atmosphere, among many other variations.
>
> In the case of mankind, many changes will be necessary so that we manage to adjust ourselves to the transformations

which are occurring in this moment of transition between HTimes. The Planicio Terráqueo is continuing in its development. And in spite of the fact that these modifications can cause intense disturbances in us, we should understand that they're part of a process of evolution which exists in all of the Fusion. For example, it may seem that the variations in activity in the sun cause us damage, but they are part of these necessary changes.

We are in the process of trying to enter into the Fourth HTime, but we need to achieve the level of comprehension sufficient to be able to do so. If we don't reach it, we'll be left behind in relation to the development of the Planicio Terráqueo, repeating the lessons of karma, since we've lagged behind by entangling ourselves in contrary reasons.

For this reason, great physical, mental, and social modifications will come. All of the established order that doesn't function will have to change so that the new arrangement can express itself and we shouldn't try to stop it or avoid it. We need to remember what we are: Egos in preparation. We are not our matter. This is ephemeral and circumstantial. Our experiences are temporary; they're lessons in order to arrive at a goal and then pass on to another reason.

We need to remember why it was necessary to send us to the Earth and to inhabit a three-dimensional body which we didn't have before. The Earth is a beautiful place, but it's a place in which we are isolated and limited, until we understand why we rebelled against the laws of our development. Our evolution here is about the awakening of our konsciousness by means of the lessons programmed in each one of the Seven HTimes. If we are able to comprehend them, our konsciousness will evolve and will allow us to perceive more fully the Reality of ourselves, our matter, and the energies and Kosmic reasons that surround us in each instant. We will perceive in a different manner, we'll once again perceive more through sound than through vision, and with this, the manner in which we communicate with each other will change. We are programmed to develop 23 senses, not just the five we have now.

The language of the future will be perceived in a multiple dimension based on the intonation of the key-words that man will use, and will emphasize something that provokes a mental exaltation, for some very great, for others, incomprehensible, according to their evolution. Greater sensitivity, greater comprehension. Thus, words will be like keys to doors that open the mind of man to a different tonal,* to great and new spaces of understanding, and another environment of greater and deeper reasons will be seen. It will be spoken inwardly and the words will be like mantrams. (*Tonal: the sonorous reasons that define the characteristics of each HTime. Each of the Seven HTimes is a different sound.)

What marks man will be his evolution. One will not speak, certain sounds will be emitted and thus, man will adjust himself to his primary and secondary reasons, according to his urge for evolution, and time will not be lost. This will bring about the establishment of differences in evolutionary powers. Words will provoke sensory stimuli, not memories of trivial things, and what will be transmitted by thought will no longer be coarse images, rather something more sublime. The arts, painting, literature will return, but they will be symbols different from the current ones.

The events of the future cannot be specified exactly since they are retracted or flavored with the reasons that the pequeño lives (26 Sep. 87). What will happen will be affected by our advances or setbacks.

Moreover, huge transformations in our matter are coming, because when the Ego evolves, our matter transforms itself. Our development will be a constant transformation, as much in our konsciousness as in our matter.

Matter is a wealth of powers that in the future will lack the variants that cause pain, sickness, aging, etcetera. It will be a pinnacle and then a transformation in which its energy will be concentrated, until it "disappears" forming a halo of Lux. The apparent decline is in reality a transformation.

Man in the future will comprehend that beauty is not one prototype, they are many. Matter will be in harmony: all of its parts in unison, to convert itself in a plenitude of Lux; the reasons of liver, pancreas, heart, etcetera that currently do not function in unison will be fused (12 Nov. 88).

We can see the apparent changes in the matter of the pequeño and we say that they may appear astonishing, but the Ego itself has suffered some minimal changes since its arrival at the Planicio Terráqueo. In the course of the HTimes the Ego will again have the appearance of a flame of Lux, which is its true exencia (essence) *since that is what it is, sonorous Lux (27 Jan. 05).*

Our matter will again be as it was before, the layers of the Ego will change until we have the form of a flame of Lux and we project our sonorousness with the completeness and grace that we had before in the Kosmo of the Trinak.

A total and absolute mutation is being developed that can take centuries, and we are at its beginning (2 Jul. 88).

The duration of the Third HTime and each one that follows will depend on our understanding and actions. As we pass our lessons, the HTimes will be shortened, but if not, it will be like the case of children in elementary school who when they haven't learned their lessons, need to repeat them over and over until they do. We can only return to the Kosmo of the Trinak when all of the Egos have completed the lessons that fall to us on the Earth and thus, we will be ready to move on to other lessons. This is the task and responsibility that each Ego has.

When we speak of returning to the Kosmo of the Trinak, we should understand that this is not a physical act of transporting ourselves there, and even less of leaving in spacecraft, but it will be the result of our kosmic development.

In this process of evolution, the Ego will finally accept and will understand what true greatness is; we will understand that it's not the mistaken urge of wanting to expand ourselves in order to take over everything, *but the inner development of understanding (4 Nov. 87).* Thus we will enter into another reason in our path of evolution, in order to one day Realize ourselves within the magnificence of the Fusion or Creative Expansion.

And how can we arrive at our destiny? By means of the elevation that comes through Kosmic evolution. We already know that we aren't going to get there if we don't make the effort to understand. As a first step we should accept that our evolution is the most important thing, and the lessons are unique and specific for each of us. We will leave behind the bonds that we've formed and the karmic reasons that trap us in a cycle of constant repetition. Thus we will be able to start out on the road that will lead us to relating ourselves in harmony with our reasons, with the Konceptos, and with the Voluntad.

This is the hope that is offered to us with the delivery of the Konocimiento of this Third HTime.

Just as in the material life, the Ego introduced into matter passes through different periods, thus, when the Ego is necessarily detached from the same it will Project itself, in a zeal for improvement that will last throughout the entire time of the Fusion, we can therefore say that its trajectory is marvelous and incommensurable, it will pass through many levels and many advances, all of them marvelous, amazing, according to its efforts, to its dedication and above all to its Submission, since comprehending and this is vital, the Submission will not be other than the harmony within it and its own desire, this will determine its entry into a new Projection of which, the current human mind cannot understand. This, if it is understood, will explain the greatness which that magnificent Beginning offers us, and it will make us comprehend that it is not necessary to have so much confusion in this state of being that is just one day, thus we should understand it, that one day are the multiple incarnations, and a few hours the current experience; our sorrows and our mistakes are part of a primary education in which we are placed to understand and to evolve and so, humanity should know this in order to trust and adapt itself to that reason which in this moment is delivered as an Absolute Truth (4 Nov. 07).

239

Helena closed her notebook and looked outside, observing the people walking in the street below, absorbed in their errands, unconscious of all of the kosmic magnificence surrounding them. When she thought about the future transformations that would come not only in the Ego, but in all of the Fusion, Helena wished that man would recognize himself again as part of that great family which is the Universe.

Chapter 15
The Protentos

Man wants to conquer the Universe to manipulate
and make slaves of those that he finds in his path.
"Arrogance," not knowing that the slave is he,
because of his miniscule figure.

SENTENCE

The sound of the firecrackers that announced the arrival of the Fiesta of the Locos resonated inside Helena's apartment, making the windows rattle. This was a festival that was celebrated every year and the town took advantage to go wild. The people headed out into the streets to shout and celebrate, infected by the idea that they could disguise themselves however they wished without being criticized. As Helena had jokingly said, "They're releasing the *chamuco* (devil) within." Diego and Max washed the dishes in the kitchen while the other members of the group arranged the chairs in their places after the hours of study they had had with Helena.

Max was so fed up with the gossip of the oh-so-peculiar owner of the House of the Coyota that he commented, "When I find Xóchitl spying again, I'm going to throw the scorpion that I trapped today at her. I'll toss it under the door when she's snooping around."

"No way, did you really keep it, man?" Diego asked him.

"Oh, yeah, I forgot to tell you; it's in a bottle on the bathroom shelf." Max chuckled, thinking about it.

"I don't want to miss that, but the poor scorpion! If it stings her, for sure it'll be the one who dies and she won't even notice," Helena exclaimed and they all laughed, ridding themselves of the tensions that overwhelmed them.

In that instant Helena perceived a less haughty Max, even though she knew that he was still battling to control his arrogance. Max had overcome, up to a certain point, his rages, even though he was still hurt by the lack of recognition that he had received before in France. Helena had caught him several times attracting the attention of the others or trying to show off his intelligence in small details. It worried Helena to think that this would leave him vulnerable and that his energy would be diverted again towards the reasons of his past.

When they finished, Max and Diego said goodbye, closing the door behind them. Helena stepped out on to her balcony for a moment. A short while later she heard Xóchitl's screams rising above the noise of the street below her, followed by the echoing sounds of her slipper striking blows wildly in the hallway. Helena chuckled and thought, 'it looks like the Coyota couldn't stop nosing around even for one night…' She smiled and lit a cigarette to reflect upon the latest events they had lived.

Helena thought about Diego and the powerful lessons which he had experienced in the last few months. His recovery after the attack was slow. Although he was progressing very well, he still suffered from a weakness in his right leg due to the energetic attack. The energy of the desiderato had affected his nervous system. Now, Diego was living a new lesson in which he'd have to stop losing himself in the environment and manage to pull his strength together. His relationship with Laurita continued, but it became more and more complicated over time. Diego suffered from a very old indolence; he didn't want to be responsible for himself and he was always looking for a relationship which he could put his energy into in order to push the transcendent things to one side.

But apart from Max and Diego's reasons, there was something that worried Helena even more. In the last few weeks when all of the disciples had gathered together to study, Pedro had come up with many excuses not to attend, saying that he was too busy with his job. The few times that he had been present, he made a point of showing off to highlight his importance. But the most significant thing was that he wasn't giving importance to the reasons they needed to study, and when the missions they would have in other places were spoken of, he turned a deaf ear. He tried to impose his ideas and the others tended to follow him, since he used his intelligence and charisma to enfold them. Pedro had the responsibility to teach them, just like Helena, but he treated them more like his friends than his students. Helena observed that in spite of the years that some of them had been studying with him, Pedro hadn't helped them to really understand the studies or to comprehend their deepest reasons, or the bonds that existed among them. If they didn't achieve a complete understanding of these reasons, they wouldn't be able to advance within the Konocimiento Kósmico. Helena could observe the confusion that emerged among them with respect to what they needed to do in San Miguel, and that they were reluctant to accept that the site was only a point of meeting and preparation from which they must depart to other places in order to teach and fulfill new tasks.

Helena noted that underneath it all, Pedro's actions were preventing what must be: a submission and union between all of them within the teachings. This union was what would allow them to cleanse themselves of their confusions, pacts, and personal interests, so that the Konocimiento would reach all of humanity. She saw them as if each one were part of a body that should function in harmony, but in which all of the parts were acting in discord. Helena only wanted everyone to find that adjustment which would help them to move forward. She had explained to them that there shouldn't be friendships or alliances within the studies, as these affinities could lead them to confusions in their actions or to activate negative programmings that were latent within them.

Yes, there were many things to study and analyze. With a sigh, she rose, closed the balcony door and went over to her desk. She picked up her notepad and continued writing the history of mankind that she would narrate in her book.

A few days later, Helena was leaving her apartment, and when she opened the door she saw Diego standing there, with a frightened look on his face. She invited him in, and when they were seated he told her in a disheartened voice, "Laurita is pregnant and she wants me to marry her."

When Helena heard this, she made a great effort to control her distress and replied, "Well, I imagine that you must be very happy, you've gotten what you wanted. You're in love with her, and with this latest turn of events, you'll be even happier. Her family will welcome you with open arms."

Diego listened to her words with pain and bitterness. After the attack which had left him weakened and with a scar on his cheek, he felt fear, so he put himself to studying his reasons. He made an effort to understand how he had gotten to that point and he realized how he himself had woven that tangled mess. When he had gone out with many women, he was receiving a great deal of energy. This made him feel self-satisfied at being able to have it all: on the one hand he had the studies, which he used to give himself an importance and to take energy, and on the other, the material world he believed himself capable of controlling. But now his game had fallen to pieces and he was trapped once again in a commitment. Diego observed his arrogance and thought, 'Why did I start playing with things if I knew they weren't going to end well?' He remembered Helena's words: "As long as we don't know who we really are and what motivates us internally, we will be our own worst enemies." She had many times emphasized the point to them, saying, "Know yourself, for there you will find the greatest truths."

Diego pulled himself together and replied, "I know, Helena. I hope I've hit bottom. I've been recognizing many of my faults. I talked to Laura again, I invited her again to the studies, but she says they don't interest her, although she'll let me do what I want."

"Sure, as long as you maintain her, take her out, and care for the baby, right?"

Diego lowered his head. From whatever angle he analyzed it, it was very clear that things weren't going to work out.

Helena remarked, "I don't believe that I can help you any more, Diego. You've studied and have advanced a few steps on this road. Up till now the Konocimiento has been given to us as if it were candies, but the effort that my Maestra made to elevate herself in order to understand it, and thus deliver it to us; well, no human being has any idea of the suffering that it cost her. I hope that you appreciate the labors of the one who has helped us so much. You know what you have to do, so I'll leave you so that you ask." With this Helena entered her bedroom and started to work on her book, leaving Diego alone so he could find his truth.

Diego took time to study and ask. After a while, he approached Helena to talk to her about what he had comprehended. "I understand that we see the act of having a child as a personal reason, I mean, as if it belonged to the parents. But we should realize that these are circumstantial reasons in which a specific Ego arrives so that, as much for the child as for the parents, we fulfill a karmic or spiritual reason. The Kosmic reasons are deeper and aren't related to our predicaments. I wasn't thinking of anything but the moment. Ok," he smiled sadly, "more like I wasn't thinking at all. I also understood that in the future the relationship between parents and children will be different, but I didn't really grasp how."

"Yes, I just now found this study; I'll read it to you." Helena entered her bedroom and picked up one of her notebooks. When she returned, she read to him: *"'The reasons that thou hast called*
genetic, and which prevail in a continuum in the actions of matter, were not as they are now, nor as they will be in the future [...]. In the future, due to imponderable Kosmic reasons, the genetic action will be deeper, to the degree that in the child there will be an almost absolute frequency of the mother, in a near totality of structures that will acquire a great modification' (12 Jul. 96). Thus, if the mother has a certain percentage of evolution, the child that corresponds to her will also have the same level of evolution. This is so they can help each other in their development. You'll see that the act of having a child is more complex than what we think."

Diego remained silent, thinking about what Helena was marking for him and of how hard it was to understand and overcome his deficiencies. He remembered what his Maestra Helena had told him many times: "Negativity is a lack of comprehension."

Lost in his thoughts, Max walked slowly towards the bookstore where Pedro was working. He adjusted his sunglasses and looked up at the sky, wanting to see if there was any possibility that it might rain that day in order to calm the intense heat of summer. While he strolled along the cobblestone streets, he reflected on how this relationship with Pedro had transpired. Since their first meeting almost fifteen months earlier, Max had felt an affinity with him. Pedro was one of the few people in San Miguel with whom he could talk about the subjects that still interested him: history, culture, technology and more. Some weeks earlier, Pedro asked for his help with a novel that he was writing. Max felt flattered that Pedro recognized his intellectual abilities. They spent their time discussing the weak points of the book and talking about history. Max felt the pleasure of having his mind occupied in something that would showcase his intelligence again. He couldn't talk to Diego about the subjects that interested him since he considered that Diego lacked preparation.

And Helena was busy with all of the things she needed to do: the studies, her book, healing, and her efforts to teach the group. This task hadn't been easy for Helena, given that they were all people with very strong personalities and since they didn't have the preparation that Pedro should have provided them, they weren't uniting together. And now Helena was talking about going somewhere else. Thinking about all that this implied unnerved him. Max had supposed that after a period of time in Mexico he was going to be able to return to France, but now it looked as if it wasn't going to happen. He was beginning to think he'd never see his native country again, or at least not for several years. He felt the need to find a little bit of stability and peace in his life. He arrived at the bookstore and went in to start working with Pedro.

Recalling the theories that her Maestra Marion had taught her and which had so astonished her when she heard them for the first time, Helena spoke to her disciples about the formation of the Fusion.

"Imagine that the Fusion, what we know as the Universe, is like a great bubble in which are contained the Kosmic realizations: those we see and those we don't see; the Kosmos, the suns, the galaxies, the atoms, the rocks, or the Egos. Everything has a characteristic in common: it is formed by the magnificence of Supreme Entities, which we will call Protentos. Through their activity, they allow all of the reasons within Life to be interrelated and develop with a purpose and in harmony. Making a simple comparison, we can say that they are the 'ingredients' with which the Kosmic realizations are formed. The Protentos *[...] by common consent join together their activities 'structure and grace' and in this way they begin to strengthen and put into movement [...]* the Fusion, promoting the Kosmic reasons. *The Kosmic is like an animated structure. The Protentual gives it the grace and the purpose (2005).* It's through this brotherhood of grace that everything within the Principle of Life can Realize itself.

The Fusion or Creative Expansion is evolving and in a constant transformation. What we see and perceive today will be very different in the future. *The Fusion will continue forming itself in millions of centuries and this is said thusly, for the incidental understanding of the pequeño (2005).* These magnificent Protentos allow it to continue transforming itself, activating or diminishing certain angles of it.

Within the Fusion millions of Kosmos exist, which are fields of development. Each one is formed with different parameters and characteristics, and take in a multitude of reasons and particles which experience their own evolution.

All of the Kosmic structures that develop themselves in the Fusion *are bathed by the splendor of these Protentos*, although in different proportions, which gives them a multiple of possibilities and variants.

"The Protentos not only formed the Ego, but *are also forming us intimately in our matter, they mark our destiny, and they give us certain characteristics. Some characteristics given by the Protentos are latent within us and will be awakened in the measure in which we go about having a greater evolution (Nov. 10).* For this reason, there is no Ego equal to another, nor one matter equal to another, since each one of the Protentos has granted its graces in different intensities, and these manifest themselves according to the development that each one of us has."

"You've told us that the Lux is the principal reason that makes up the Ego. Then, are there other reasons?" asked someone from the group.

"I'll try to explain it simply. We can say that *we were made through the coexistence of the Protentos, although in the same way as a cake is mostly flour, we belong more to the reasons of one of them, the Koncepto of Lux (Light), than to those of the others (19 Sep. 87).* Remember that *we are Kosmic structures, delivered to be within the reasons of the Lux in principle and saturated by the other Protentos (2005).* The Protento Lux has different manifestations of itself, each one with a specific function. We belong to the Lux Sonora, Lux that converts itself into sound. We are a thinking sound. For this reason our projection (thoughts, emotions, words, the actions that we carry out) is sonorous. But it's by means of the actions of all of the Protentos that we achieve a plenitude as Kosmic particles, or Realize ourselves."

Attentive, the fifteen persons who formed the group jotted down in their notebooks the study that Helena dictated to them. When they got to this point they took a break; Max got up to serve himself more coffee. He looked out the kitchen window that faced the internal garden of the house and observed the sky saturated with stars. He reflected on how amazing the things Helena was teaching them were. He checked his watch: it was almost 7:30; the performance wouldn't start till nine. He went back to the others, offering to serve them more coffee, which they accepted.

He observed Diego's face to see how the scar was doing. Amazed, he saw that it had healed very well with only Helena's emanation. The words that she had spoken that night returned to resonate in him. "When are you going to realize the magnificence of what you're learning?" This phrase shook his beliefs and it seemed that he finally began to accept the fact that he didn't know everything. He intuited that he was barely beginning to glimpse the transcendence and greatness of the studies and the reality of his small existence. Recognizing this hurt him quite a bit and he continued struggling to yield, searching inside himself for a little humility in the presence of the studies.

When he had left France fifteen months earlier he never imagined that his life and his way of seeing it would change so much. He was learning more about matter and healing than he could have believed possible; those things simply didn't fit into his previous world. Since the incident with Diego, he began to cure with emanation, following the guidance of the Konceptos. It was a very ample topic for study and he immersed himself in it, desirous of understanding as much as he could. Max found a satisfaction that he hadn't felt since he worked at the Institute in France. He felt that at last he had found his way, and thought himself capable of beginning to decide things for himself without having to bother Helena with his concerns. He liked that people would seek him out to ask for his help, and the truth was that he enjoyed the attention he was receiving in exchange for his labors. He realized that one part of him still sought recognition and that he wanted to feel important. But he didn't see much danger in it if he maintained himself alert and made an effort to control it. He didn't see how it could affect anyone else.

Helena paused and turned to look at him as he returned with the beverages, saying with a smile, "Just coffee? This is going to be really boring. Speaking of cakes, I bought one from Diego's friend. Do you all want some?"

Diego blushed, thinking about his most recent experiences. To escape from the shame he felt, he went to the kitchen and returned with a piece of cake for everyone. A little while back he had broken up definitively with Laura. He had explained to her that he wasn't going to commit himself to her or to the baby that was on the way, and she had told him to get lost when he tried to explain one more time about the importance of the studies and the transcendence that they had. It had hurt Diego deeply to see how she had rejected her opportunity, and how she also was denying it to the child. Diego hoped that one day both of them would understand the transcendence of the Konocimiento and join the studies; but if not, he shouldn't have any more contact with them.

When they had finished eating their cake, Helena continued. "For the time being I'm only going to talk about some of the Protentos, since due to their 'closeness' they are the ones that we can try to understand.

The names that I'm going to give you are only a reference to their Real names, since each word expresses a power upon mentioning them. When you all have more time in the studies their true names will be given to you.

Let's continue with the **Protento Lux** (Light), with which we are intimately related. It forms the completeness of our being. The Lux is operating in all of the Fusion. *The Lux is the Protento which activates the other Protentos to mobility. It is like the spring of a watch, in this case, of all of the established (88).*

Man of the past [...] was taught [...]that the first thing that he should do was seek his inner Lux. This remembrance has remained in the cubicles of reason, and it is said, that the inner Lux is to hit the mark and comprehend, but they did not understand that the Lux was the Kommunication with the bestowed Kosmic reasons (10 Nov. 95). That's why man always relates the search for his spiritual part with the Lux, because he intuits that with the Lux he'll be able to understand his reasons.

A Lux that has a greater hierarchy is always teaching, each particle of Lux is always learning (1 May 90).

We can see this characteristic reflected in the arrangement that we Egos have always had. Those of greater development have the responsibility to teach the rest, and at the same time they must continue on in an apprenticeship.

To conclude, we will say that the Protento *Lux delivers harmony* to man *(14 Oct. 96).*

"Another of the Protentos is **Konciencia** (Konsciousness). *Konsciousness is the capacity of everything that exists to comprehend itself and to comprehend the environment (Nov. 10).* It is by means of this Protento that we can understand ourselves and say 'I am, I exist.'

Each corpuscle [...] of matter or of energy [...] has a particle of Konciencia [...]. Each proton, each molecule, each reason of the tree or of a leaf, like the feather of a bird and the bird itself, the stone, or the animal[...], man, the sphere in which thou travelest, each one is konscious, and each element, the water, the fire [...] all has Konciencia (16 Jun. 94).

In the same way, thunder, lightning, the clouds, a comet, an asteroid, everything that surrounds us has konsciousness, even though it may not be a konsciousness that we can comprehend. There are many different types and levels of konsciousness. The konsciousness of an Ego is not equal to that of an Omnipotencia, or this to that of a Creator. It all goes according to the level of development that each has. We Egos are beginners in our evolution, and this evolution is about the development of our konsciousness. Among the Egos, each one has a different variant of konsciousness, that is, each of us is in a different point of growth."

Helena continued, reading: *"'Thou shouldst understand that everything that enters into Life comprehends and perceives, and once more we insist, it is the angles of projection that gives the konsciousness and the modifications of the same'"* (21 Apr. 98).

"Is it through Konciencia that everything is related among itself?" Max asked her.

"In this study we see that, *'Every particle has the vital konsciousness of the Yo (I), this with the goal of understanding itself and evolving. A plant, a stone, everything has the konsciousness of the Yo.*

An atom has konsciousness of itself, as well as the particles that constitute it, called by man protons, neutrons, etcetera. The Ego is a Kosmic atom and therefore, it and its parts have konsciousness of the Yo. This is part of the development of a particle within the Principle Life. Everything is connected among itself and this allows, for example, that a tree feels when another tree is being burned, even though they both are separated by a distance of kilometers' (14 Aug. 01). So yes, it's in part due to this Protento that all is connected, but the other Protentos also have an influence in this interaction of reasons."

To give them an example of how Konciencia manifests itself in life, Helena read them this study: *"'There are different levels, or variants, of konsciousness, which are manifested throughout the life of a pequeño, in different moments and in diverse ways.*

A first "natural" adjustment occurs in the moment of birth. If the adjustment does not happen, the child dies. Afterwards, these adjustments may occur in a natural manner, with each cycle of

renovation that the Ego and its own matter, suffers, each seven, eight, or nine years, just as they may also occur due to a great sudden emotion that jolts the Ego, or in the moment of a grave illness.

These moments of adjustment among the variants occur occasionally, and if the Ego takes them for itself as a moment of understanding, then they will set the standard by which other adjustments happen, the Ego reaching little by little a greater fullness, until its konsciousness is complete and it even achieves transforming its matter' (17 Jun. 94).

"The actions of Konciencia are multiple in each animated 'being.' (17 Sep. 99). In the case of man, we can appreciate this multiplicity of konsciousness in our matter; for example, each corpuscle, organ, and system has its own konsciousness, just like each of the nine layers of our matter has its own awareness and comprehension. All of these konsciousnesses should be in synchrony, but most of the time they're not. The Ego is always perceiving in multiple levels of konsciousness, even though we don't realize it.

We should understand that we have a poorly developed konsciousness, and in addition, it is limited to the perception of a Fantasy. We can't perceive the Real or Reality (the Real is the beginning, where everything comes from). We were limited upon arriving to the Planicio Terráqueo when we were introduced into a three-dimensional matter, and then we suffered other limitations due to the errors we have committed and which have perturbed us mentally.

Throughout the ages, the Ego has developed the comprehension of itself by means of Konciencia, the same way as it also has utilized it in order to understand and relate to the environment; nevertheless, now man perceives more by way of his feelings and emotions and not by way of a Kosmic reasoning, which makes him perceive everything through a reflected konsciousness, making use of the environment for his recreation, thus it is that this only allows him to perceive the Reflejo (Reflection)[...] and not the Real *(11 Oct. 95).* Currently, humanity is becoming more and more attached to a perception of the present and of the moment; we have been reducing our comprehension of the environment in which we live.

But we need to remember that we only perceive a part of what is. We're surrounded by energies of the Earth, of the Genios, of the sun, and of the Kosmo, among many others. This is what our growth is about; as we develop our konsciousness, we're going to be able to perceive more of these reasons. In each one of the Seven HTimes there is an advance planned for our konsciousness, if we're able to take the necessary steps in our evolution. The Ego should allow itself to perceive other environments, not just the material one that we have accepted as reality. *For example, a painter perceives colors that others do not, a poet sees images that others do not perceive, and this perception is the result of an effort (17 Mar. 90).* In a very distant past, the Maestro permitted our konsciousness to perceive form and color. In other words, he helped us to amplify our awareness so we could perceive these reasons which up to that point were incomprehensible to us. If we make the effort and pass the lessons that fall to us, we can have the konsciousness of memory and understand the steps we have taken in this sphere. In the future man will have the multiplicity of konsciousness. The memories will be complete in us; it won't be like a fuzzy image or *déjà vu*, but like exact memories. If we persist a bit more and have the necessary humility and respect, perhaps we will be able to perceive and communicate once again with the Entities that regulate the Planicio Terráqueo, as it was in the past. At the end of the Seven HTimes, if we fulfill all of our lessons, we will have the full konsciousness of ourselves as Egos, of the sequence of our Ansibirs and of the Reality that exists here in the Planicio Terráqueo."

Diego recalled what Helena had told him in one of their first talks. *"Our quest has as its objective the development of humanity by means of the study and analysis of the Kosmic reasons (11 Jun. 88).* These studies are for developing the mind, to awaken the angles of Konciencia that are latent within us; of course this can only be achieved if the pequeño has a submission to the lessons that correspond to his moment."

The delicate scent of a flower called *hueledenoche* reached Helena and she allowed herself to be distracted by it for a moment. It was a very sweet and penetrating fragrance. Something she liked about San

Miguel was the assortment of plants and how well they grew there. There was a nursery on the outskirts of town where they sold flowers and foliage from very diverse regions. Her apartment was filled with many varieties of plants, but Helena always found space for a new one. The week before she had 'adopted' a miniature rose bush with delicate scarlet blooms and she had situated it on the narrow balcony of her apartment where it could receive the morning sunlight. Focusing again on her studies, she adjusted herself in her chair and continued with the lesson.

"Another of the Protentos is what for now we'll call '**Mod**,' and first of all we'll say that, *the Protento Mod, is simply what we call movement, since everything is in movement from the most inferior particle up to the great galaxies (Nov. 10).*

This Protento governs cycles and internal movements, but not the movement of passage from one place to another. In the case of man's matter, for example, it governs the internal activity and the routes within the body, of the mind, bloodstream, neuronal activity, thought. Imagine the quantity of movements that the body must carry out for its functioning.

The Ego also has movements which take the form of an 'infinito' (infinity). We can visualize it as a spiral that expands outwards from a center to then contract back upon itself and return to its point of origin. For example, *[...]when a pequeño wishes to record images from his environment, and to do it uses a video camera, which collects the images that later it will project in a television. This process makes the pequeño have to carry out an 'internal' movement to perceive and another 'external' movement, to understand the perceived (17 Oct. 97).* During the hours of the day we pass constantly through these movements."

Helena thought for a bit and added, "Another example is when we're in a wakeful state; we're constantly collecting from the environment and when we sleep we make a recounting of it, an assimilation of the perceived.

We see the action of this Protento in cycles, for example, menstruation, the rotation of the Earth, or in the change of the seasons. Or we can appreciate its graces in a dancer or in an athlete.

Well, speaking of dancers, it seems that it's time for us to get going. I'll go get my shawl," Helena announced, as she checked her watch.

The others arranged the chairs in their places and picked up their notebooks. Those who didn't want to go said their goodbyes, while the rest prepared to leave. The group of fifteen dwindled to eight, and they walked over to the Angela Peralta Theater. Max's girlfriend, Ninetta, had gotten tickets so that they could see an Indian dance troupe.

They arrived on time and settled into the box seats. Ninetta was already waiting for them. She was 32 years old, thin and fragile-looking, but with a strong character. Her long honey-colored hair was pulled back and her large eyes shone with delight when she saw them arrive. They greeted each other and sat in their places. Ninetta had met Max in the school of Bellas Artes where she taught ballet classes and he taught French. Max had chatted with her over the course of the months about some of the things that he was learning. After a while they began to go out and the relationship had developed little by little. Max took it slowly, since he had seen what was happening with Diego,

but when he had asked about it by means of the Kommunication, he understood that there was a reason for him to get closer to Ninetta, and this would reveal itself shortly.

One day she came down with a bad case of flu and didn't go to work. Max stopped by her house to see how she was, and in that moment he offered to heal her. When Ninetta realized she felt better almost immediately, she asked Max to explain how he had done it.

When Max had presented her to Helena, Ninetta was astounded by the things Helena spoke of. Shortly thereafter, Ninetta asked for the opportunity to study. As she was within the same path of evolution as he, Max knew that the energies and efforts of both should be directed towards the same reason of evolution, but he also knew the relationship could change according to the Voluntad (Will). He thought of Diego and the problems he had with Laurita, due to the divergent reasons that directed their lives.

The Angela Peralta was a very old theater and the wooden seats resembled instruments of torture that had survived the country's war of independence. Nevertheless, the stage was decorated in a very exotic way that night and the odor of incense perfumed the space. The theater was full, and Helena took out her fan as the heat upstairs was intense. Almost immediately they lowered the lights and the music of sitars and other traditional instruments began. The dancers emerged out of the darkness behind the curtain onto the stage, illuminated by the soft light that made the golden accessories of their costumes sparkle. Their gentle and harmonious movements trapped the attention of the spectators. Diego was absorbed watching the undulating balance of the body, which was accompanied by the sinuous movements of the hands and feet. The dancers' eyes were in complicity with their bodies, and they directed their gazes to follow the movements of their hands. Max took out his binoculars and handed them to Diego. That way he could appreciate more fully the cadence the dancers followed with the music. The musicians who were arranged on the stage floor made the strings of their instruments sing. Helena observed how the movements, the rhythms, and the melodies had an energetic purpose; this was a very ancient race that had had a moment of splendor and glory in the teachings of the first eras.

They left the theater a little bit hungry and went strolling towards the square in the center of town. Helena walked over to the cart of a street vendor who sold hot dogs and ordered two for each of them. Max was accustomed to complain when it came to eating on the street, but he didn't tonight; he accepted this act of Helena and enjoyed it without getting all fussy about it, as she often said to him.

Max and Ninetta bid goodnight to the rest and walked back to the house where she lived. Max started to talk to her about his progress on the book Pedro was writing. She listened to him and asked, "And what does Pedro want from you, Max?"

"Well, that I help him with his book, of course. Today he asked me to assist him with his research in the historical archives in Querétaro. There aren't many people here who can help him like I can. And you know, I'm really enjoying the research."

Ninetta insisted, "No, Max, he wants something else."

Max became annoyed and replied in a high-handed way, "Ah, so you think that I only deserve to be a poor French teacher? Maybe you don't understand, but I was very well-known and respected in France. And Pedro sees that I have a well-trained mind that's capable of assisting him."

"What's going on with you, Max? You know this is your weak spot. Don't you see that he's using you for something? I don't understand what he wants yet, but I don't trust him."

They walked in silence towards the house, each of them absorbed in their thoughts.

Chapter 16
Revelations

*Each human is subdued within a conformation of
programmed ideas, of programmed acts, of programmed reasons
within the same Apparent; it is like a cupola or sphere where
all of the manifestations of life move and rebound,
and from which no one dares to leave
(13 Feb. 95).*

Taking advantage of their days off during the town festivities, they all decided to take a trip to the city of Querétaro, situated just a few kilometers from San Miguel de Allende. There they meandered around the city center, looking at the Fountain of the Dogs, the historic buildings, and the walkways that surrounded the museum where they went to see an art exhibit. Afterwards, they ate lunch in a restaurant on the main plaza and finally they enjoyed home-made ice cream being sold from a cart in front of the square. They sat in the shade under the trees to savor their ices. Helena began the conversation, commenting, "See how the different trees have specific forms in their physical features which reflect the distinct functions they carry out. Their structure and their activity are in accordance with Kosmic Laws that dictate these reasons. Each particle within the Fusion is subject to Laws that mark its development and dictate its action according to the programming for which it was formed. We're going to talk about the Protento that delivers these parameters; we'll call it **Solvencia**.

The Protento Solvencia *delivers what we know as codes or Laws, since everything is arranged within a reason and a vital principle (Nov. 10). It is like an energy which promotes that the reasons of*

the Akto Protentual (Protentual Akt) are carried out; it gives it the endeavor, and accepts or rejects the actions that are not coordinated to the whole and are considered temporary (30 Oct. 96).

Every particle in the Fusion was manifested with a purpose and for a determined end; therefore, there are Laws which regulate its actions. These Laws dictate how and what its endeavors should be so it arrives at an improvement. As a simple example, we'll see that for a musician to be able to develop as such, he needs classes, practice, music theory, harmony, etcetera, and not biology or politics. In this way, the Protento Solvencia promotes the development of each creation, attracting what is appropriate for it and rejecting that which is not.

But what happens when a particle behaves outside of the parameters of its activity? There has to be period of adjustment so that it reestablishes itself in its efforts. In a way that it's put into a special education until it readjusts itself in accordance with what was planned.

The Planicio Terráqueo has its own Laws of development: triangulation and multiplication, which we have spoken about before, are part of its codes, but there are many more.

Within this Protento are the Laws which help define the different periods, adjustments, and lessons through which the Ego has passed and will pass in its stay in the Planicio Terráqueo. We'll see that in stone, as well as in the animal and then in man, the Ego has received the parameters necessary for its development. The conformation of our matter, the processes of Trebolo and Ansibir, wakefulness and dreaming, the form of our perception and projection, everything is made in accordance with specific Laws that give us an adjustment and an accommodation."

"And the laws that man has for his coexistence, where do they belong?" someone asked her.

"Morals and the norms of coexistence among men are human reasons that man has retaken from what he has been able to perceive. *Each moment of evolution has its tendencies which are resolved in ideologies of conduct, these are modified erroneously until they are inappropriate or they thrive correctly until reaching a plenitude.*

This is decided by means of the libre albedrio (free will) that the human race shares, and on an incomprehensible scale (21 Jun. 92).

These norms of conduct, just like morals, are interpretations that man has given to the perception he has had of the Protento Solvencia. Many times these discernments haven't been understood fully or have been subjected to the interests of groups that have used them for their convenience, resulting in great delays in the development of humanity. For example, man fixed a code of human conduct which favored a hierarchical structure and the interests of just a few, but not to promote an adequate spiritual development. These human laws don't explain anything to us about how we should relate with each other or with the Konceptos that are in the Planicio Terráqueo and in the Kosmo; nor did they tell us how we should seek our evolution.

Whether or not we know the Kosmic Laws, we are subject to them. To know them implies understanding how we should behave, and for what end; they also indicate where we're going to, and why. We'll understand the lesson that we should have in its correct moment. Knowing and fulfilling the laws will give man Freedom."

At that moment a group of children who were running in the plaza set off some firecrackers that exploded close by with a deafening noise. The doves congregated on the ground flew away frightened, leaving them alone under the trees of the plaza. Diego felt a shiver and his body shuddered. He said to Helena, "Wow, how strange. Suddenly I saw a battle scene. You're not going to believe it but I even smelled the gunpowder and the blood on the battlefield. It was just a flash, only an instant and it was gone…"

"Yes, a sound can activate the record that we have of our previous life experiences," Helena remarked.

"And how is it activated, Helena?"

"In the Espíritu Huella the **Protento X** manifests itself. *Thy matter has in itself the X, which is that which endures and gathers all of the actions and reasons that have revolved on itself (26 Mar. 98).* Just as man's matter stores this record, the Planicio Terráqueo also has a memory of all that has occurred. *Each pequeño has a special and unique X that is identified with sound (2005).* The Ego *is a kosmic atom in preparation and transformation, introduced into the*

reasons of sound, into the 'Lux Sonora' (Sonorous Light). It contains a special filament that represents its 'history,' since its beginning until its end. This we have called Espíritu Huella. It is in reality a halo of energetic Lux and gives sonority of tone, which makes it unique (2005).

This Protento manifests itself in the stone, animal, or man in the layer of matter that is called Espíritu Huella. It is by means of the Espíritu Huella that the Ego receives the information stored in the X of all of its actions on the Earth and since before arriving to it. In it are the habits and the graces of all of the eras, what we are and what we will be. That's why we have memories in us, and in a given moment they can come to mind."

"And why was it that memory and not another one, Helena?"

"Ah, well the Espíritu Huella has its own time for manifesting itself. It is there when we can see how the other Protentos that we've talked about take part in this apparently simple reason. Nothing happens by chance; there is a reason and an order.

"Ok, this leads us now to talk about the **Protento HTiempo** (HTime). This Protento gives us the position within the Fusion. *[...] Everything has a Time and we are within the HTiempo (HTime) of the Fusion, which concurs with the HTiempo in the Kosmo Jehovah. The HTiempo is points of reference (16 Jan. 09).* HTiempo is not a measurement as we have understood it up to this point, nor is it linear. *[...]It gives the calculation of possibilities to everything which the Fusion represents (Nov. 10).*

HTiempo acts as a container in which are situated the reasons that correspond to the development of a particle. These reasons present themselves in a manner and in a sequence which HTiempo defines, in concordance with the other Protentos. As an example, it situates a person in a specific moment in which the other reasons that are necessary for an experience are going to converge.

HTiempo (HTime) has been defined to thee in different interpretations, and therefore, the measurement of it is not the same in all environments (1 Jul. 93). For example, time on the Earth is not equivalent to time on the sun. In the same way, the Ego Trebolario does not experience time in the same way as an Ego in the Ansibir, or an Ego in dreaming. An Ego in stone does not perceive it in the same way as an animal or a man. HTiempo gives the location and the points of reference to the particles so they can develop their actions. In the case of the Ego, it lives in one time relative to its projection in matter, another relative to the Planicio Terráqueo, another to the Kosmo and yet another to the Fusion. All of these times are perceived in a distinct way."

"But why are there so many times?" Max asked her.

"In order to understand this proliferation, we'll say that, *[...] HTiempo (HTime) has very diverse variants, being itself only one, and that it projects itself in a multiple which is incomprehensible for ye (23 May 01). We will understand that HTiempo is one. From HTiempo depart the Times as thou hast understood. The Times gravitate in the same HTiempo, therefore there is only one transcendence in the present, the past, and the future. We will give, as an example, that thou hast a bookcase, it is as if in that bookcase we would fill it with books and we would take the book that pleases us, but in the same bookcase were all of the books that shall be necessary (28 Feb. 92).* Imagine that the bookcase is the Protento HTiempo and each book is a different Time. Inside of each book are many other 261 reasons that join together in a harmonious way to make up a story. Each book gives a different purpose and situation. We perceive the past, present, and future as separate measurements, but for example, a greater Entity perceives all of the Times in the same instant." Helena noticed Max's surprise.

"Is it really like that?" Max asked.

"Yes, we're living the present, the past, and the future all at the same time, but we don't have the capacity to be able perceive it like that. In other words, what you live right now is the result of your past actions, and what you do in the present will define your future. Then we can say that the future comes out of the past. I'll explain it to all of you. In the case of mankind in the Planicio Terráqueo, we say that we're in a Tiempo Peremne (Peremne Time), that is, *it is a Time that is delayed, even though the relationship of events goes forwards in another dimension, in another relationship or reason of Times. Everything that we live is the repercussion of a Time that already was.* That's why we understand things with a certain delay.

For example: an explosion, the repercussion of a deed that already was reaches us later on. This is because we are disconnected from the events of the Fusion since we are behind in our apprenticeship. *As we advance we move closer to a Tiempo Exacto* (Exact Time). *The Tiempo Peremne is further away from the primary reason,* from our Real reason *(Carpeta Dorada).* The Ego in the Planicio Terráqueo lives Time in a repetitive manner in which all of the events repeat themselves until we understand."

Recalling a study, Helena commented to them, *"In the course of its presence, the Planicio Terráqueo has not had one of the forms of HTiempo, but we could say, various reasons of the same [...]. These reasons were not adjusted to the measurement of thy clocks, and that yesterday, today and tomorrow cannot be measured by man. This relaxation of wanting to measure time is a game that is granted to the pequeños, to give an idea of something that in truth has no comparison (21 Sep. 96).*

When we speak of 30 million centuries ago, we do so according to our perception and possibly it has not been 'so much time.' That is why it is said that everything that is lived is a repeated time, the past is lived and at the same time the future. Thus man has to live Seven HTimes, in other words, he has to live the same thing seven times in a different way, in different times, and the consummation of each HTime is comprehending (27 Feb. 88).

As I mentioned before, we perceive Time in different ways, depending on the awareness of the Ego in each moment. For example, in small variants, when we feel a great heartache or worry Time runs slowly, and on the other hand, when we have felt a great happiness, we say that Time flew. This is an example of how HTiempo interacts with the Protento Konciencia, delivering variants. That's why we perceive Time in different ways, such as playful, changeable, or infinite. There are variants of HTiempo that we aren't konscious of.

"Now we know that there are a multiplicity of variants of HTiempo, Lux, Konciencia, Mod, Solvencia and X. What we perceive and sense is in accordance with these variants, that is to say, with the possibilities that they deliver to us for the experiences that we must live. What one person must live is unique, and in the same way, what one person perceives and understands is not the same as for someone else.

Ok, until now we've talked about some of the Protentos. They've been given other names and small references have been made to them regarding the manner in which they manifest themselves in man. But we should remember the incommensurable transcendence that they have in the Fusion. These reasons, for the time being, cannot be perceived by us until we have a greater konsciousness, product of our development. It remains for the future to continue on in the studies of these magnificent reasons that form and promote Life. They're studies that we'll develop in other moments."

"And in a manner of speaking, are all of them in us?" Max asked, who as always, had an interest in understanding the function of matter.

"The Protentos are in harmony in us. This study explains to us: *'So that thou understandest, if all of the notes on a piano are played at the same time, this causes a profusion of inadequate and inharmonious sounds. But if the notes are played according to a defined order, then a melody is obtained. In the same way the Konceptos Protentuales (Protentual Koncepts) manifest themselves,* 263 *some in a specific moment, and others in another. This is the result of that equilibrium that thou callest harmony. We can say that the Konceptos Protentuales (Protentual Koncepts) form a melody of the nine bodies that make up man in the present, and also it is clarified for thee that this melody is affected by the causes of the pequeños. These causes, thou shouldst understand, are all that which provokes an alteration in the equilibrium and the development. This of course is not dependent upon the social or moral causes, since these are an invention of man (7 Feb. 01).*

There are many reasons for which matter has all of its characteristics [...]. Sometimes the Ego can request to have more of one Protento than another, according to the reasons or activities of the Ego in matter. For example, if a woman wants to be a dancer, she gathers more of the Protento Mod, or if she wants to be president, she gathers more of another Protento, such as Solvencia [...]. In the same way, when an Ego is in the Trebolo, it can request that it be given certain excellences, and can persist in this until receiving them, as

for example having beauty, although it is always granted only that which it deserves, according to a previous pattern, and so if it does not fall to it to have a maximum beauty, it will receive the most beauty that it should have according to its merit.

The idea of the Protentual is that it gives an improvement, delimitation or rejection'" (13 Jul. 97). With this last explanation Helena finished the lesson.

Appreciating the beauty of the city upon the finishing the class, they heard the voice of a man relating stories about what had happened in the city hundreds of years ago drown out the murmurs of the birds that arrived each afternoon. Helena turned to watch a streetcar filled with tourists who were taking pictures of the plaza while they listened to the narration of the driver.

"Mmm. The coconut ice cream was delicious," remarked Helena.

"You can't really want another one?" Max asked, knowing how much Helena enjoyed that dessert.

"Oh, of course not; what can you possibly be thinking? Well, ok, but I'd like melon this time."

Once they arrived in San Miguel, Max went to the bookstore and greeted Pedro, who was behind the display counter. Pedro indicated to him that they should sit in the two chairs located in front of the shelves filled with books. When they were both seated, Pedro opened a folder filled with papers and checked the observations marked on the first pages in red.

"I was reading your annotations in the manuscript and you have some good ideas, Max. I knew you were the right person to help me with this project."

Max perked up when he heard these words and responded, "I enjoy delving into those intricacies of history. If I hadn't studied medicine, I would have liked to teach history."

"Well, it shows." Pedro made himself comfortable in his chair and took out his pipe. He carefully packed it with tobacco and lit it, filling the shop with aromatic smoke. He took a moment before starting to speak, searching for each one of the words that he wanted to say. "You know, Max, I've noted that you're very apt for this road that we've chosen in the studies. You're extremely intelligent, but also intuitive; it's a marvelous combination. You have a lot of strength to carry things to completion. You know how to analyze and I see that you've advanced a great deal in the studies."

Max smiled to himself, wrapped up in Pedro's flattery.

"Very well, I wanted to speak to you about something, Max. You know that Helena is obsessed with the idea of going to another place. But I also have my own plans. I'm considering giving workshops in Oaxaca in order to have something permanent. Some of the people from the group are thinking of coming with me. It would be wonderful; we can teach the studies, heal, and there are so many other things we can do. Do you know how many people would follow us? Why don't you come with us Max? You and your girlfriend, we're going to need you. Imagine everything that we can do together. You with your great mind and your dedication, I tell you, there's no limit to what you can achieve."

Max began to play with the idea in his mind, contemplating what he longed for. 'Umm, it wouldn't be bad to have a moment of peace,' he thought. But at the same time he felt uneasy when he crossed looks with Pedro. He saw something in it that made him feel uncomfortable with himself. A warning ran through his body and he recalled Ninetta's words: "What does he want from you?" Max shrank back. He realized that Pedro had involved him in his games and that he had gotten to him through his weak spot: the need to receive recognition for his intelligence and abilities. Max felt clumsy in the face of this manipulation of Pedro's; his pride was wounded at having fallen into his trap. Max interrupted him abruptly, with the excuse that he had forgotten about an appointment and that he needed to go immediately. Max left the bookstore feeling extremely uneasy with himself; on the one hand because he realized what his deepest motives were and on the other for having been caught up in the plans which Pedro had hidden from Helena.

That night Max had a dream that upset him deeply. He saw himself walking towards Helena's apartment and, entering stealthily into her bedroom, he spied her emerald ring on the night table next to her bed. Silently, he took it and left, running through the streets of San Miguel until he arrived at Pedro's house. He found Pedro standing in the doorway, waiting for him. Max moved closer to him and delivered Helena's ring. He saw the look of malice on Pedro's face and in that moment he awoke, sweating and breathing rapidly from the fright.

He got up to write down the dream and began to study it. What he understood left him stunned. He comprehended that he had known Pedro in the distant past and had a programming and a pact together with him. In that time, Max had assisted Pedro so that he could establish himself as the maestro. Pedro had used the Konocimiento to promote his interests, while Max exploited the studies in search of a material immortality. Max understood that in the present he was aiding Pedro again, delivering his energy to him. Pedro was taking it to repeat what he had done in the past: separate himself from the group and destroy it so that the Konocimiento would not arrive in plenitude to humanity. Max had been participating in this error without recognizing it, but in reality he was fulfilling an old agreement. He felt pain and shame when he saw his own reasons and that murky part inside himself that he still had not finished cleansing. He decided that he would go look for Helena the following day in order to tell her what he had understood in that moment. He didn't want anything to damage what she, with so much determination, was carrying out.

Chapter 17
The Division

Do you want to know who the bad guys are?
It's very easy, they're the ones who want to govern the rest
because they believe themselves superior.

SURPRISE

Seated on the balcony of her apartment, Helena drank her coffee sip by sip while she contemplated the experiences she had lived in the year and a half she had been in San Miguel. She was distracted from her thoughts when she noticed some grey clouds covering the sky, warning of a coming storm. Helena adjusted her sweater while the breeze lifted the dust from the street to form a small whirlwind that passed rapidly in front of the building.

She had had great hopes upon returning to Mexico and finding herself with Pedro for being able to fulfill what her Maestra had always yearned for: to form a group dedicated to the Kosmic studies. This was the reason for which she, as much as Pedro and the people who accompanied them, had come to San Miguel. They were all pieces of a puzzle; each one should find his place and affinity within the studies, so that together they could form a solid foundation that would allow them to develop the Konocimiento of this HTime. But then the problems with Max and Diego appeared, along with many others, and now it seemed to Helena that Pedro was facing some very difficult internal tests that were affecting them all. San Miguel was a point of reunion in order to then move on together to other places, but Pedro hadn't accepted this reason. The experiences in this location had been intense, since the site acted like a mirror in which each of them could see their past experiences reflected in their present actions. Helena pondered the importance of understanding these reasons of the past that are latent within us. 'Sometimes we act, thinking we are doing the right thing, but the result ends up being a mistake. It's because we don't have the clarity to know what it is that propels us to make these decisions. We act more from habits that are recorded in us than according to what we should do.'

She understood that Pedro was letting himself be carried along by reasons of the past which he hadn't wanted to see. He was distancing himself from his commitments with the studies and putting his interest and energy into the bookstore where he worked. There he spent his time, distracted by the women that he attracted with his intelligence and humor, but more than anything, with the force that he received through the studies. He looked bored, absent, and showed resentment towards Helena on the rare occasions they got together to study, since he usually said he didn't have time. It seemed that he was only interested in finding out what studies their Maestra had left Helena, without putting his effort into understanding and developing those that he already had. He just wanted to have more and more, like a child who can never get enough sweets. Helena sensed Pedro's intentions, which were to separate himself from the formal reason of study to continue on by himself. If he did so, he was going to take the force that study had given him to establish an antagonistic reason since this formed part of his negative programming. Throughout his incarnations he always sought to have the command, installing himself as the maestro and making the decision to deliver only the teachings that he thought appropriate for the rest. Thus, Pedro maintained a control over the teachings in order to have a power and dominion. Helena wished they both could surmount the tests they were facing in these moments so they could overcome these lacks of understanding. She only hoped that Pedro wasn't really going to leave, like Max said he was planning to do.

Humanity has always been in an internal battle between the Lux (Light) and darkness, angles and demons, good and evil; this is a theme as ancient as man himself. It is spoken of in almost all cultures and the earliest stories describe it, since this memory has remained recorded in the register of the Ego and of the Planicio Terráqueo itself. Since the Egos arrived at the Planicio Terráqueo, they were offered the opportunity to understand and correct their error, which in the beginning was not that severe. Some accepted the conditions of the apprenticeship, but others did not, and they divided themselves into two bands. The opposing band has searched for the way to alter our matter and the mind of man throughout the ages, since with these acts they take away from humanity the possibility of harmonizing themselves with the Kosmic reasons and thus perceiving their truths. The battle between the two groups is becoming increasingly more present now in this moment of transformation and adjustments in the Planicio Terráqueo.

Studying and analyzing world events, Helena was certain that the current intention of the leaders of the rebellious Huestes was to focus the energy and mind of man towards this battle, without people realizing it. They managed it in a very subtle way. The world culture was saturated with images of a chaotic future, with destructions and a humanity that lived in deterioration and on the brink of extinction. This idea had been introduced insidiously in everyone's mind through movies, books, games, even with the interpretation of the prophecies about 2012. They were obstinate in the idea of destruction and violence, incapable of imagining a world that didn't revolve around material possessions and power.

She thought, 'Another way they alter the mind of man is by filling it with fears: of terrorism, bankruptcy, illness, of growing old and dying, among many other things, and thus they maintain everyone in a constant state of disruption. But where they're most determined is with young people, since they know that they have the strength to make changes. They've got the minds of the youth absent and imprisoned in videogames, drugs, or the internet. The idea of globalization is part of the same thing. The leaders of the world want to unify the thoughts, tastes, and ideas of everyone. But each race has its reason, karma and apprenticeship, the same as each Ego, which is unique and individual. Each one is at a different level of development. That's why it's incredibly negative to try to equalize all of humanity in the same thought.

The only way to break out of this mental confinement is by searching for our truths. If we don't do it, we're only living according to ideas that they project as our reality. We know that this is a falsehood, but we haven't understood how to remove ourselves from it or where to direct our efforts. The Konocimiento of this Third HTime will give man the power to break with these impositions.'

A hummingbird flew over to her, loudly protesting that its feeder was empty. Helena was always surprised that such a tiny animal could make such an uproar. She smiled and got up to fill the bottle of sugared water.

She returned to her thoughts. 'When our konsciousness was limited to the awareness of the physical world, we stopped perceiving how energy functioned in other dimensions. We're no longer konscious of the way our energy is projected or taken back in, nor of how it's being manipulated. We don't see how the energetic bonds function between us. The leaders of the world take this energy of the masses for their purposes. Their goal is to make this illusory world into which we are introduced even stronger, so 270 the Ego doesn't perceive its truths. When we fall into their games, we are actively participating with them in making the walls of our mental prison more solid and it's increasingly more difficult to escape from it.'

For that reason her Maestra had always indicated to her that each human being should be individual, and therefore it was vital to cut the bonds that we have with all of these mental structures, which are represented principally in the commitments that we form with a country, a family, or a religion. Religion is the return to the origin, Hueste (Host), legion, or group to which we belong since the Kosmo of the Trinak in order to comprehend the error that came to fruition based upon these groupings. But now we have to overcome the idea of religion, since it's a closed circle which constantly repeats old and already surpassed ideas. It's negative because it doesn't change or advance. It only encloses man in rites and beliefs that limit his development and understanding.

'How different would it be,' she analyzed, 'if we knew that this fantasy which we live in is a human construction designed to divert our energy and stop our Kosmic development; thus we'd stop suffering for an absurdity, for something that isn't Real.'

To cheer herself up, she took out one of the last studies that her Maestra had delivered to her regarding the possibilities for the future, and read: '*The Konocimiento manifests itself in different ways. Some will not be understood, exactly, or partially, by thee pequeños, but we could say, hypothetically, that they are like syllables, or words or phrases that do not have a direct or complete meaning, but when they are united by a directive act, they themselves can give an explanation and complete manifestation. Such is the case of the Planicio Terráqueo. We could say that it is like an encyclopedia, and that each individual and each plant are a small phrase, or a poem, or a small story. When the Konocimiento contracts within itself, it forms a harmonic expression. Thus it is that the Konocimiento affirms itself, like when the writer narrates a sketch and manifests it with breadth. But as the writer changes a narration in a decisive way, to open the path for another that in spite of being different, later is united with one and then another enunciated expressions, so it is how a compilation of Konocimiento disappears in appearance, but later resumes and continues the storyline. This is what ye consider as incarnation.*

[...] It is thus that the Konocimiento, for thy small understanding, forms a plant, a stone, an animal, a man or a star. All of this is manifested in a singular way, and it is thy konsciousness which limits the complete form, to accommodate one that is adjusted to thy perception, which is limited for now, since the brain is, to thy understanding, a machine or a reason that simplifies the forms, according to its own norms and limitations, the same ones that, with determination and submission, can focus themselves towards other reaches and other diverse perceptions, since the Ego was designed for it, but this will be when the Ego comprehends its own grace, and utilizes it for its well-being and evolution' (28 Nov. 97).

She remained in silence for a moment and inwardly delivered herself to the Konceptos, asking for the strength and clarity in order to continue and complete all that they would ask of her. She got up to start getting ready and then went out to look for Max and Ninetta.

Once more, Xóchitl checked the interior of the enormous house to make sure no one was there. The owner of the House of the Coyota slowly opened the front door, and the brilliant sunlight fell over her dark brown face, blinding her for a few instants. She raised her hand to protect her eyes from the radiance. When she saw the tall man waiting on the other side of the street, she waved so that he would come over quickly and whispered to him under her breath, "It's a good thing you got here so quickly, señor. I don't know how long they'll be out, you'll have to hurry."

Pedro adjusted his hat to cover his face and took a moment to check the street, assuring himself that no one was watching him. He quickly entered the building, carrying a backpack over his shoulder. Xóchitl immediately closed the door behind him and bolted it; in any case she would remain by the entrance to stand guard. She took a key out of her apron pocket and with a twisted grin on her face, handed it to Pedro. Pedro took it and surveyed the form of the key in the palm of his hand, hearing the beating of his heart in his temples. He tried to calm himself, raised his eyes, and taking the steps three at a time, ran upstairs like a bloodhound on the scent. When he got to the door of Helena's apartment, he doubted momentarily what he had come to do. He breathed deeply, and controlling the slight shaking of his hand, put the key into the lock and turned it. When it opened, he felt he had transgressed something far beyond a door.

The clarity of the light which bathed the interior of Helena's apartment stopped him for a moment. All of his body was tense and it seemed to Pedro that his mind was playing tricks on him. Small blocks of images from other times appeared before him; they were flashes that, although he was unable to precisely make them out, left him with a feeling of discomfort. Something very powerful inside of him shouted at him to stop. He told himself not to pay attention to it and that he should search for the papers quickly. He entered the bedroom and opened a closet where he found a drawer filled with papers and folders. He checked them and found the file that he wanted, removing it and placing it in his backpack. He ran out of the apartment leaving various papers strewn on the floor.

He passed by Xóchitl, who with an ample, malevolent smile on her face, was waiting for him to unlock the door. He delivered the key to her, but when she tried to shake his hand, he rejected her and said in a contemptuous tone of voice, "I don't know you; we've never seen each other before."

Xóchitl was stunned; she had expected Pedro to be grateful for her help, but she hadn't seen anything but disdain in his look. She walked across the central patio, and when she passed in front of her beloved macaws, they started to shriek and make a ruckus. It seemed as though they were reproaching her for something. She went to the rear of the building where her apartment was situated to look for something to eat. She only wanted to get rid of the uncomfortable sensation arising in her abdomen.

Diego began to feel dizzy and a tension in his body put him on alert. Uneasy, he left work to go search for Helena, excusing himself by saying that he had very bad stomach pains. All morning long he had powerful thoughts and feelings of rage and anger against Pedro; surely it was because he was doing something despicable. Diego began walking as fast as he could, given the weakness that he still felt in his leg, and arrived at the apartments with some delay. When he went upstairs, he saw that Helena's door was open. He entered cautiously into the living room, not knowing if he was going to find an intruder there. He heard a noise in the bedroom and leaned in, ready to grab him. But to his surprise, he found Helena checking her files.

Helena had already realized that Pedro had taken the folder with the last studies that Marion had left her and which she herself had been analyzing and developing. When she saw Diego, she said with certainty and fury in her voice, "It was Pedro. He stole one of my folders. I came running home when I felt something was happening, but I got here too late. I found papers scattered on the floor and several studies are missing."

"Bastard! I think that's what Max's dream was about. How's it possible that this jerk has gone so far? Let's go look for him."

When Xóchitl heard Helena and Diego arrive, she decided to investigate what was happening. She walked in front of the apartment, with the excuse of watering the plants that were in the hallway. She saw the furious look on Helena's face and several thoughts passed rapidly through her head. 'Damn, they found out so fast; a little earlier and they would've caught me. I'd better go to the back patio and hide; for sure they're going to come and ask me what happened.' Anxious that they not see her, she slunk downstairs. But when she got to her apartment, she tripped on the step of the entrance and fell headfirst to the floor. Her head struck the stone doorframe, breaking her neck instantly. There she remained, alone and out of everyone's sight.

Helena, unaware of the situation with Xóchitl, and wrapped up in her desire to recover her studies, closed her apartment door and together with Diego went walking as quickly as possible in search of Pedro, anxious to know if they would find him still at home or if he had already fled.

274 A sensation of security and willpower guided Diego's steps; he was surprised that he didn't feel his weakness. His thoughts bounced from one thing to another, but he was able to analyze what had happened since they had arrived in San Miguel. He felt a huge admiration for the effort Helena had demonstrated. Diego didn't understand how Pedro could have acted so despicably, knowing the transcendence of the studies.

Helena felt a hole in her stomach from her desperation. They arrived at the corner of Umaran street and turned right; they were very close to the house of that charlatan Pedro. They were relieved to see that his car was parked in front. They passed through the front gate and crossing the small patio, reached the front door. It was open, and Diego stopped in the doorway to block the exit. She said to him, "Wait for me here. Don't let him escape with the studies," while she entered in search of Pedro.

The interior of the house was dark and the contrast with the sunshine outside blinded Diego for a moment. When his eyes adjusted themselves to the dim illumination inside, he was able to distinguish that there were two packed suitcases in the living room and several folders on the coffee table. Diego wondered if these were the papers that Pedro had stolen from Helena's house.

Two tall, well-dressed women came out of the kitchen when they heard someone enter the house. They asked themselves, 'who is this woman and what is she doing coming into Pedro's house?' One of them tried to stop Helena by questioning about her presence there; but Helena, with all the confidence which characterized her and the anger she felt in that moment, walked past them without answering and went directly over to the table where the folders were lying. One of the women went to advise Pedro to come out of his bedroom while the other drew nearer to Helena, trying to stop her. However, she shrank back when she saw the look on her face.

"I only came to take back what was stolen from me. If you want to know more ask Pedro why he did it." Helena calmly picked up the documents and began to leaf through them. They were the ones she had worked on in Paris. Her Maestra Marion had entrusted her with studying and developing what transpired in the Second HTime, while Pedro in Mexico should have completed the other part: that is, what had occurred in the first eras. Thus, knowing what happened in the past, they would avoid repeating the same thing in the present.

Pedro rapidly exited his bedroom with another suitcase in his hand. When he saw Helena standing in front of the table and Diego in the doorway, he tried to disguise the alarm that he felt. The two women stood by his side to support him. The tallest one, gesticulating with her hands, yelled at Helena to get out of the house, saying that Pedro was no thief. Diego remained in the threshold, waiting like Helena had told him. He knew that Pedro was going to try to flee with the studies and he would do everything possible to stop him.

Helena's heart beat forcefully in her chest. When she caught Pedro's eyes, she met with a cold, hard stare. Pedro had stopped posing as he always did and she was perceiving an image of him that she hadn't seen before. In middle of the silence that surrounded them, she heard the first raindrops fall.

"Tell me, Pedro, why did you do it? Are you going to run like a simple thief?" She looked at him fixedly, incredulous at seeing that he was planning on slipping out of town like a common burglar.

"I'm leaving and some members of the group are going to follow me. I have a wealth of opportunities far from this place. They're going to help me get established." He indicated the two women by his side. "Ariane and Lucy have numerous contacts with money who can assist us. With their help, I can gather together more people and teach. I want to make my own road, and I don't need you for that. I'm more than able to do it." The two women smiled, showing their support. Pedro continued. "I'm a maestro and I'm vastly limited here with you." In a haughty tone he added, "You know that people follow me. Or haven't you realized it?" With a smile that resembled a grimace he directed his gaze to the two women, who were moving towards the table with the intention of taking the folders.

Helena was stunned in the face of Pedro's audacity. She said, "Are you putting together a harem to worship you or a group for the studies? Is it because of this foolishness that you're going to lose the greatest thing there is? Don't fool yourself, Pedro. The ability you have to attract people isn't yours. Have you forgotten so quickly? It was given to you so that you could fulfill the first part of what you had to do: gather together a group. If you go, you'll no longer be in the studies, and the grace that was given to you will be taken away."

Outside, the rain began to fall more heavily and the lightning illuminated the living room with a brilliant white light. The peals of thunder resounded one after another; it seemed to Helena that they were falling right over the house. Ariane and Lucy stopped walking towards the table, amazed and frightened by the loud noises that resonated inside the building and shook the walls. Pedro thought of lunging at Helena to get her out of the way so that he could grab the studies, but he saw that Diego didn't move from the doorway. He would have to try to dominate Helena in another way. Pedro noticed that his strength increased with the energy he was receiving from his allies; he thought that with this he would have enough force to confront Helena. They both maintained their gaze and Helena felt a change in her perception; she entered into a different time in which everything moved rapidly. She felt that a powerful energy was spinning around inside the living room, wrapping her up in it as if it were a small whirlwind. She knew that Pedro was lancing his energy, attacking her with the rage that he felt. She protected herself with emanation.

Suddenly a flash of lightning lit up the house and Helena observed that Pedro's face had changed. She saw herself in a different place, in the central patio of an ancient dwelling filled with palm trees and with an octagonal fountain in the center. She noted stone walls, columns, marble floors, and furniture typical of a very archaic house. A hot, dry breeze brushed her face and she understood that she was in the middle of a desert landscape. The touch of a light cloth caressed her skin and she caught sight of herself dressed in a long, diaphanous garment made of a thin fabric that was soft like silk. She realized that she was in the same scene she had seen when she was a child, nine years old and on the point of drowning in the sea. Once more she was arguing with the man she had seen in that vision. He was furious, and he was shouting that he was the one who should be in command and that he knew how to continue. Helena noticed there were other people standing around the patio; they were arguing among themselves. Some supported the man and others supported her. The man wanted to leave and she was trying to stop him but he didn't pay attention to her. Suddenly he lunged and grabbed her forcefully by the hand, pulling her, making the emerald ring that she was wearing fall to the ground. Helena noticed that the man in the vision had the same eyes as Pedro, and although they didn't resemble each other physically, she knew with certainty that it was he. In that instant the scene faded away as quickly as it had arrived. Helena was once again in the present, in Pedro's house in San Miguel.

She observed Pedro standing in front of her; he hadn't moved. He had a strange look on his face, as if he had seen a ghost. He stared fixedly at the emerald ring that Helena had on her hand. When he looked at the jewel she could see the rebelliousness of centuries summed up in Pedro's eyes. They were filled with greed and the longing for power. She understood that he had seen the vision, and he had recognized himself and her as well, but he did not feel any remorse for what he saw of his earlier actions. On the contrary, he insisted on repeating the action that had been his ruin in the past. The storm outside intensified and the claps of thunder made the house vibrate as if these were arms carrying the property and shaking it.

"Marion gave me the studies and I can do what I want with them. And you aren't going to stop me again," shouted Pedro.

Helena felt the blow of the energy which encircled her, trying to confuse her and take away her strength. She knew that Pedro, together with Ariane and Lucy, were attacking her. She perceived filaments of lux that passed between the three of them, and understood that the women were fueling Pedro with their thoughts and emotions. She saw that Pedro's appearance changed, his body transformed itself, and Helena perceived him as a dark shadow which absorbed that energy. The shadow grew and it seemed to Helena that he launched filaments of energy that buzzed and flew like birds around her. Pedro tried to touch Helena to shove her out of his way. With all of her power, she emanated a white lux that filled the space and broke the connection between him and the two women.

Diego remained standing in the threshold, stunned, perceiving the confrontation on the astral level and supporting Helena with his energy.

Pedro's face darkened with the fury he felt and he screamed, "It's always been you who's prevented me from rising to the top. I don't understand why they gave the ring to you and not to me. I deserve to have the same as you. Well now I'm going to do what I want. Don't you understand who I am, who I've always been? Since the beginning, I've always been within the studies."

"Yes, I know very well who you are. You've forever wanted to put yourself at the pinnacle, trying to control and direct the rest. Your failing has always been the error of mankind: not understanding the humility and submission to their Creators. Since the beginning, you've used the teachings you received to have a mental control over the people who followed you. Your intentions have been to confuse the mind of men, manipulating the teachings and delivering only half-truths, and a half-truth is worse than a lie. With that you've taken away their chance to advance. You've done this over and over again. You haven't wanted to accept that we're here to overcome our failings. You've received many chances to fulfill your responsibilities with the Konocimiento throughout the HTimes Pedro. If you already remembered what you did before, then understand all of the consequences which that act had. There won't be any more opportunities for you. When you leave you'll never again have clarity to understand the Konocimiento, nor to teach. Give me the studies; yours and the ones you stole from me. You won't be able to elevate yourself; you won't have the permission to have the studies in your possession. With your arrogance you've rejected the Voluntad and you've betrayed the Maestro who gave the Konocimiento to humanity. We're nothing Pedro, only a tiny part of the Fusion. Why do you insist on dominating and having power?"

"Well, you know that I have the studies in here," he said, pointing at his head, "but if you want, take the folders. With what I know, it'll be enough. Get out of here and leave me in peace."

Helena couldn't believe that Pedro could have gone so far. With the rage that she felt because of his insults and offenses towards the studies and the Konceptos, she indicated forcefully to him, "You alone have just broken the relationship with the Konceptos, and with that you marked your sentence, like in the past in the Kosmo of the Trinak. And you'll carry the guilt of your Hueste because it's an error of *lesa humanidad* (crime against humanity). In other words, because of your fault, all of mankind will experience a deterioration of konsciousness. And all of humanity will point to you as the culprit of this regression. That's why many people become konscious of a rejection of a person or of a specific ethnic group, because deep inside they sense that these people are part of the error for which they are being punished. The Konceptos withdraw themselves from you, as you have offended them with your actions. You will no longer have the Kommunication nor the abilities that were granted to you. You must never again touch a study, neither in this lifetime nor in any other, and much less, teach. If you do, in that instant you will lose your lucidity, but not completely; there will be a remnant, enough for you to understand why you're like that and the damage you caused to all of mankind. The elevation you have is not taken away, but never again will you have the opportunity to return, just like those who left in the past. You have offended everything and everyone. This can be forgiven but what will not be pardoned is you having offended the greatest there is, that is, the Kommunication with the Konceptos, what man has called the arrival of the Spirit in him; this you will remember always and you will never pardon yourself for it."

"Get out of here! I don't believe anything you're telling me." In spite of the menace in the words Pedro hurled at her, Helena saw the fear in his eyes. Ariane and Lucy kept on supporting him and encouraging him with their words. Pedro continued, "I don't need you; I won't listen to another word from you. You and that idiot get out of my house in this instant!" Pedro grabbed the files and threw them at Helena. Some of them fell on the floor and Diego ran over to pick them up and brush them off, astonished by Pedro's cynicism. He felt such anger in the face of this offense that he wanted to hit Pedro.

Helena became enraged at this lack of respect and contempt Pedro had just demonstrated before the greatest there is. She gathered up all of her force and marked, "You are no more than a devious, underhanded Indian who accepted possession, and with deceit you have convinced the others. You will never find peace, for all of the centuries of the Seven HTimes to come. I don't want to remain any longer in the presence of the greatest traitor there is."

Helena and Diego gathered the folders of studies and wrapped them in his jacket to protect them from the rain. They walked out of the house together.

Diego had observed the confrontation between Helena and Pedro in silence, astonished at seeing what, in reality, happened. He remarked to Helena, "Wow, Helena, I can't believe that idiot did what he did. Who does he think he is? After so many years in the studies, to lose himself like that in an instant."

"As you've seen, it's very easy to lose sight of the objective. The Continuo and our own negative reasons are incredibly strong and are always pulling us. We're surrounded by and subject to interests, as much external as our own internal ones, that don't want us to arrive at our Realization, because if we do, they lose all of their power. We need to tread carefully. Remember that being in the studies is like walking along a double-edged blade; there's no room for error. The only way to find a balance is asking and fulfilling what we are requested to do, to the letter. It's as my Maestra always taught me."

"And what are you going to do to Pedro?"

"Me? I don't have to do anything; Diego. I only leave him to the Voluntad of the Konceptos."

It was the last time they saw Pedro. As they headed towards Helena's house, the rain stopped and they continued walking in silence and thinking about the forthcoming repercussions of what had happened. They knew that when an initiate fails, great catastrophes come in all of the dimensional planes, in addition to the physical one.

Helena knew when they didn't accept and didn't want to fulfill the Voluntad, that Pedro, as well as those who had previously left the studies, broke the link with the Konceptos. One of the things that was asked of them was to hand over their notebooks of studies as a symbolic reason so that they would not touch them again, and to thus avoid that they continue speaking about them, since without having the sanction of the Konceptos they would only confuse the people who heard them, as had happened in the past. When they no longer had the Kommunication with the Konceptos they would not have the clarity necessary to be able to teach, and this was the manner in which they would corrupt the truth, as had already occurred before. They would probably lose the right to have matter and they would remain left behind in the depths of the stone. And if they would one day arrive at having matter again, they would always carry a mark and would feel an enormous emptiness. All of this was indicated by the Laws of the Konceptos.

When she arrived at her apartment, Helena went outside to sit on the balcony, reflecting on all that had happened. She was exhausted and very sad. Her eyes filled with tears and she felt a profound sense of desperation. She hoped Pedro and the others who left with him would not corrupt the studies they stored within themselves, nor deviate the truths she needed to transmit, as had always happened. Helena was beginning to understand the disillusionment that her Maestra had felt all those years ago in Palenque when she saw her titanic efforts torn to shreds in the space of a moment. Helena felt the enormous weight of the responsibility that now fell fully on her. She had never thought to be in this position. Now the all of the commitment of teaching, caring for, and comprehending the studies rested on her.

As a result of the vision and the battle with Pedro, everything became clear to Helena, confirming what she had studied in Paris. The vision she had in Pedro's house was the memory of a faraway epoch when she was within the group of initiates who had received the teachings of the Second HTime. The emerald ring was given to her in that time, as a symbol of her commitment with the Konocimiento and to signal her elevation. Pedro was also in that group of initiates which had the responsibility of delivering the teachings and the truths to mankind. When the Maestro had gone, he left them prepared and with the ability to fulfill their tasks. But the ages-old interests of power and dominion once again arose; as a result, the disciples divided themselves. Each one decided to act according to their interests and convenience, forgetting about the Voluntad. Some had not wanted to teach, keeping the studies for their own vainglory, and others formed alliances with groups of power, including politicians, in order to subjugate the rest. In that time, Pedro had wanted command over all of the disciples, and that was why he tried to take the ring from Helena. Of the teachings of that HTime, these men only delivered half-truths, storing away the rest in order to maintain the power among themselves. With these acts, humanity remained subjugated and limited. The idea was promulgated that elevation and comprehension were only for the selected few, and they could only be obtained through the hierarchy that maintained the power. With this treason they closed the possibility to mankind of being able to have the Kommunication with the Konceptos. They submerged humanity in fear and in "no-podernimiento"* (*the limitation of "I can't").

She sat there gazing into the distance and touched the mole on her forehead, contemplating what was going to ensue now. She heard someone knocking at the door.

She stood up and walked towards the entrance. She opened the door and found Diego, Ninetta, and Max standing outside. They commented, "We wanted to see if you were ok." Max examined her; he had never seen her so sad. He felt terrible at realizing how close he had come to failing as well. He saw how he had been supporting Pedro, giving him his strength. Diego and Ninetta came closer, worried about her.

With a nervous and shaky voice because of the thought of what had transpired, Ninetta said, "Diego told us what happened, Helena. It must have been terrible."

"Yes, it was very intense, and it hurts me more than I can say that they've separated from us. This shouldn't have occurred. But the eleven of us who remain are going to keep moving forward."

"What's going to happen now, Helena?" Diego asked.

"Diego, can't you see she's exhausted?" Ninetta exclaimed. "Leave her in peace for a while."

"You don't know Helena like we do. She's made of Lux and steel," Max added, looking at Helena. He smiled, trying to cheer her up.

The four of them went into the living room and made themselves comfortable. In that moment her confusions dissipated and Helena understood fully that she had the strength to continue forward. She had strived for more than twenty years to perfect her Kommunication with the Konceptos; she had confidence that they were going to guide her steps, even though she knew that her tests were going to continue in this Ansibir and in all the following ones. It would be a constant process of development, as it should be. And as she already had the sequence and comprehension of her previous lives in place and had overcome her mistaken programming, she knew that what would come for her in the future would be new lessons in order to be able to advance within the Kosmic reasons, leaving behind the reasons of karma. She looked at them and said, "Well, we're going through a very difficult moment, but we need to carry on and fight. You and the others who have decided to continue have an enormous task to fulfill, the same as I do. We need to be firm, convinced of what we're doing, each one of us and as a group. It's an uncertain future, and we always need to remember that our only refuge is the question and the humility that brings us closer to the Konocimiento."

"And what's going to become of Pedro and those who left with him?" asked Max.

"The people who have left the studies and then try to present themselves as maestros are doing an extremely grave damage, since they can't teach if they don't have the Kommunication with the Konceptos. They don't have the clarity that the Lux and the Kommunication give us. *When a pequeño takes the Konocimiento and later leaves it, he plays with it. He becomes disturbed and remains crazed throughout the centuries. Each memory that he has of the past vivifies the cause, and is the motive of a new imbalance. The great madmen are very intelligent pequeños who profaned the mysteries"* (8 Apr. 88).

Diego asked her, "Why did the others decide to leave with Pedro? Didn't they know?"

"Pedro hid many reasons which weren't convenient for him to teach them. He didn't talk to them about pacts; he always rejected this idea. And they're related to Pedro precisely because of the pacts they formed with him in the past. Pedro knew that if they understood and broke with these reasons, he would no longer have their support."

Diego was silent for a few moments and then remarked, "You know, when you were confronting Pedro, I saw other things. You know that since that demon attacked me I can perceive things that happen in the astral level. When you were both talking, I saw how that guy's energy enveloped you, like a grey cloud, trying to dominate you. I saw dark filaments leave that cloud and in one moment it seemed like birds or something were flying around the room, attacking you. I felt Pedro's anger like blows of energy thrown at my abdomen. I felt dizzy and it gave me a terrible headache. You know, Pedro's voice was affecting me too; it was like a hissing which made me sleepy. When that happened, I noticed some grey shadows were moving in the living room, uniting with Pedro to give him their energy. And he was using that force to attack you even more powerfully. Then I saw you were emanating a white Lux that expanded out like waves. It was so strong and powerful, it removed the grey mist that Pedro was projecting and it broke the flow of energy of the women and the shadows which were fueling him. It was at that point when the storm outside got stronger, and between the lightning, the sound of the thunder, and your emanation the shadows withdrew from the house, and Pedro was left alone, in front of you."

Helena, after listening to Diego and his account of what just happened, commented, "What you saw was how Pedro was receiving the energy of all the people and the forces that support him. This happens by means of the energetic bonds that have been formed in other times between humans and the Huestes of Egos which were in the depths of the Earth, and since when we clumsily released them before their time. And now they're the ones who have, among other things, the mental, economic, and political control over humanity, and will do everything possible to not lose it."

Ninetta remarked to Helena, "You know, I haven't been studying for very long with all of you; could you explain what a pact is to me?"

"Each Ego is unique and was formed with a different yearning, konsciousness, and will. For that reason each one has to live a distinct evolution. Pacts are agreements which are made between Egos in order to protect an interest. This interest has been that of not accepting that we were formed with a purpose which we need to fulfill, and that there is an order and laws to be able to carry it out. Pacts were formed in the first eras when the condition of education on the Earth was not accepted. The vital part of this education is that each Ego has to look to its own evolution. When this didn't happen, we put our energy into structures that shouldn't exist, and we go on forming more karmic ties all the time.

We are submerged in three great erroneous ideas. The first is the desire to dominate each other. The second is the search to make our matter immortal, rejecting the cycle of incarnation and denying the sequence of lessons that should form a continuity in our evolution. The last idea is the intention to flee from the Planicio Terráqueo in order to not submit to the instruction that we are asked to have. These are the desires of the great hierarchs of the Huestes; they are the groups who direct the world. Everything comes down to the fact that the Ego has still not accepted its error or found submission before the Konceptos. We continue in rebellion when we think that we are the owners of our acts, that life is a coincidence and that we can do whatever we want with it. They have made us believe that libre albedrio (free will) is being able to do what we want, but libre albedrio is to attract that which is advantageous for our development.

I'll give you a simple example of how these ideas have damaged us: we were given a beautiful place to live in, but what condition is the Earth in now? We were given a wonderful body that was nourished by its environment. We lived for contemplation and without illness. What did we do with that matter?

We are a principle of Life within the Universe, we're in development and apprenticeship, and for that reason we need to be guided by the Konceptos. Mankind on its own could not have evolved."

"And how can we move forward?" asked Diego.

"How can you help yourself? Stop delivering your energy to these three ideas that form the base of the reasons of the Continuo (Continuum) which we have already spoken of. The Continuo envelops us like a dense mass that puts us to sleep and prevents us from elevating ourselves to perceive the great Kosmic reasons. When you separate yourself from these ideas, you'll be able to begin to perceive the Kosmic reasons. Then, the Ego can find itself and its destiny. That's why it costs us so much effort to elevate ourselves, because we have to cross through that layer of contrary thoughts.

When we elevate ourselves, the rebellious Egos are exposed and they don't want their game to end. That's why they seek to keep man's mind constantly occupied, perturbed, and asleep. There's nothing more dangerous for them than a mind that thinks for itself. Above all, they know there is no way to counteract the thought that has a truth.

It's vital that all of humanity understand these reasons and are able to cut their own bonds. We within the studies have already learned this, but it's necessary that everyone understand it. We have to learn to ask permission before acting, and accept that there is a reason behind everything. We should ask about even the most trivial things, since it is there where we always get carried away by our tastes and habits, and adjust our actions. For example, ask if we should eat this or that; if we should take a medicine or not; if we should have a relationship with so-and-so. We have the Perespíritu to guide us, and we should develop this kommunication with it, and then strive in seeking a relationship with the Konceptos. If we don't do it, we're not going to be able to leave the Continuo to fulfill our destiny in the Planicio Terráqueo.

"And so it is, guys, they're going to come at us with all they've got. In regards to the studies, they're going to accuse of wanting to overthrow the social, moral, and spiritual order, and say that what we talk about are harmful ideas. But that Ego who is prepared and open will find the greatest truths in the studies. And when it recognizes them we hope that it lets the positive part of itself come out."

"And what follows for us?" Max asked her.

"For the time being we're going to continue with the plans to leave Mexico. We have labors in other places. Now, to rid ourselves of the bad taste in our mouths, I'm going to read you this study." Helena opened one of her notebooks and began to read: ***"We are in a process of cleansing in order to be the beings that were originally conceived. We are asked if we want to return to that original perfection. The recovery is difficult but the Konocimiento cannot be received without it. To achieve it, physical and psychic imbalances will come, but if we do not do it, no one will arrive. Man by himself cannot find the way, he always becomes lost because he distances himself from submission, from the guidance of the Konceptos"*** *(16 Jul. 88).*

Chapter 18
Resolution

The Universe is a small bubble in the middle of a great destiny.
When will mankind understand it?
When madness disappears from them and they
convert themselves to humility.

SENTENCE

Helena stayed in her apartment all morning. The pain she felt in her stomach had not stopped and seemed to burn her from within. She recalled what her Maestra Marion had commented to her years earlier: "When you unite yourself more fully with the Konocimiento, you'll speak just like I do; don't be amazed if you hear yourself using the same words I do. You'll also develop your strength to attract what you need and to reject everything that obstructs your path." Her Maestra had paused and stroking the corners of her mouth, as was her custom when she was about to tell a joke, added, "And you'll also have my stomach aches and headaches and the other discomforts that you don't know about. Ay, Helenita! But whatever it takes to evolve, right?" Smiling at the remembrance, Helena went to her closet and took out the last letter she had received from her several years ago and read:

Helena,

The Konceptos have told me that I must leave because it's not possible that I continue receiving more offenses on the part of humanity. You know well that I don't want to go, but if that's what is indicated to me, what can I do, in the face of the orders of the Konceptos?

When you are ready and have understood what happened in the Second HTime, you will have to return to Mexico. There you will meet Pedro, who remains here studying and trying to understand the reasons of the past of the Americas, so that together, you can face the changes that are coming. Perhaps in the moment when you return, there will be more people with him.

I do not know the reasons of the Konceptos, but they tell me that it is to you that I must entrust the care of the studies and that you must gather together all those who are involved with them. Together you will have the force to liberate humanity from the imposition to which they have been submitted, since it helped its hierarchs to escape before the required time.

I only came to deliver the Konocimiento to you all; it's the only way in which man can elevate himself and understand his reasons. The struggles and battles between men don't interest me.

Throughout all of the HTimes there have been so many people whom I have taught, and of all of them, only one has succeeded. As Omar Khayyam said, "I had one hundred pearls, but I could only thread one." I trust you will be the second and that later, many more will arrive. I hope for the day when the Egos, free of their hindrances, integrate themselves once more in the Universe.

I wish you luck and remember that when you feel alone, you will only find refuge in the studies and in the Kommunication. Be humble and little, and may what manifested you be within you.

She heard someone knocking at the door. Helena put the letter back in its place, thinking that many of the things her Maestra told her, and which in the moment she didn't understand, had manifested themselves over the years. "And all the ones still to come!" she sighed, while walking towards the entrance of her apartment to see who it was.

Max and Diego had arrived with the intention of inviting her to go out. Helena was happy to see them and thought, 'it seems they're willing to continue; they've been advancing in their development, even though who knows what other tests are coming for them. Meanwhile, one more pequeño is on the way with Ninetta's pregnancy. This baby is going to be able to grow from the start in accordance with the teachings of the Konocimiento and the Mandato (Mandate) of our evolution. I imagine a future when all of us are born already knowing who we are and what the reasons of our existence are. We'll know that the Konceptos are going to counsel us and guide us in our apprenticeship.'

Max was especially jovial. "We only have a couple of days more in this place and we thought we'd go out to celebrate. If you want, we can walk up to the lookout on the hill. The afternoon's cool and quite pleasant," he remarked to Helena.

Helena was grateful for their good intentions, and making an effort since she wasn't really in the mood, she gladly accepted the invitation. "Ok, but you're going to have to wait for me to get ready."

"No problem, I don't think they close till midnight." Diego joked.

Max smiled at Diego's joke. Seeing him fully recovered and cheerful, he thought, 'what a tough way for Diego to understand how energy functions and how we're connected to the Continuo (Continuum) and among ourselves. Well, it was a very severe way to learn, however through that process the ability to perceive events in a more real form in the Astral was opened up for him. He came out of it stronger and more focused; it's going to serve him well with everything that's coming.'

After the long walk through the steep cobblestone streets, they arrived at a restaurant which had a magnificent view from its terrace. It was a house typical of the town which had been renovated for this business. It wasn't a very well-known spot, so they could be at ease conversing and eating without the tumult of diners. Max and Diego were talking about their experiences and the concerns they had felt since childhood. Each one realized how everything in their lifetime had interwoven itself, forming the thread that led them to the encounter with their destiny. It was the small decisions, the ones that perhaps in the moment didn't seem transcendent, through which they had found the way to draw closer to the Konocimiento.

They were there a good while sharing their anecdotes until they arrived at the most recent ones. Diego joked to Max, "And now you don't even have a tiny office, Doc."

"And what happened to all of your girlfriends, Casanova?" Max replied.

"Let him get back on his feet, since his name bears his penance." They both stared at her, waiting for her to explain.

"Di-ego, *mujeri-ego, andari-ego**." (*womanizer; wanderer)

The jokes continued for a while. When they got to dessert, Diego ordered three glasses of anisette and they toasted to the next journey and place of residence. Then, he addressed himself to Helena. In these last few months Diego had observed her tenacity and the incredibly strong willpower she had to be able to gather them all together and to understand the reasons of this odd place. She had put up with many of his absurdities and his foolishness. When he told her he felt ashamed of his previous conduct, Helena replied, "It's the duality that's within us, it's in all men, we want and we don't want at the same time. It's the inner battle that we'll have incessantly until we adjust ourselves to the Voluntad and we defeat that which disturbs us. When we recognize this duality, we're looking into the face of the hidden enemy that is ourselves."

"Helena, thank you for what you've taught me. The truth is that it never occurred to me that I could one day be able to understand why we exist. Ok, and much less understand what we are in Reality, where we come from and where we should be going."

"They're the three questions that have always been in man's mind and have never before been answered," added Max.

Helena replied, "Your gratitude isn't for me. Without wishing to offend, we infinitely thank the greatest Entity that came to the Planicio Terráqueo and taught us. Through his efforts, we received the opportunity to approach the Konocimiento." Helena remained in silence for a moment, contemplating how difficult it has been for man to show respect toward the most elevated Kosmic reasons. Returning to the topic, she asked Diego, "And how would you summarize what you've learned, Di-ego?"

"I'd start by saying that I'm an Ego, not a soul, nor spirit. And that our matter is a disguise that changes in each lifetime according to what we have to live. We have to use a body as if it were our car so we can learn through it. We always receive what we need in each Ansibir to be able to advance. And, well, that we're on the Earth for being arrogant fools. That's why they locked us up here like delinquents in a jail, and it's here that we need to stay until we understand. It sounds harsh, but that's the way it is, right? And we continue on without accepting that we blew it and we go on complicating things, leaving our mind really disturbed and the Planicio Terráqueo pretty well destroyed. And damn it, on top of it all, there's the hierarchs that won't let us advance. They and their 'powers' and their craziness wrap us up in material dreams making us believe we're a big deal, so that we don't wake up." Diego, excited, paused and added, "The truth is that I want to wake up from this nightmare to someday see Reality."

"Well, for my part, I really want to meet the rest of our family and see what else our future holds in store," commented Helena, wishing humanity would understand that there was a multitude of possibilities before it, if only it would, in this moment, take the first steps towards its destiny.

Max moved his chair, angling his body away from the table. That way he could stretch out his long legs. For a moment he felt himself to be that young man again, restless and curious. Wishing to express what he felt to Helena he said, "You know, the truth is that when I accepted to stay and enter into the studies, I wasn't completely sincere; I recognized some of the things you talked to me about, but I was curious to see how far I would be able to go, I wanted to test my aptitude." Max ran his long fingers through his hair nervously; it had always been difficult for him to articulate his internal reasons. Enthused by the moment they shared, he continued. "Yes, I know, like always, I was seeking the maximum, something that would make me feel exalted and important. Now I realize that instead of advancing I was limiting myself. The desire to observe the world and dominate it was only distancing me from the true possibility of comprehending it. Science was restricting me, even though I believed the contrary. It's here in the studies where I understood that when one has the humility to grab onto the hand of the Konocimiento, that he is accepted and guided to understand what we can't even begin to conceive of as our Reality. The truth is that it's cost me a lot to be able to accept it and stop imposing my will, but I've understood that even though my lifetimes have unfolded in different settings, what has motivated me has always been the same, and what I am now is the sum of my past actions. It's like removing layers and layers of fictitious wrappings, to begin approaching my Reality."

"I, for my part," Helena said to them, "want to tell you that these studies we recently began as a group are just doors we should open through our efforts and by means of the Kommunication. Little by little we'll develop them through the Seven HTimes. The important thing is that we achieve the continuity in our Ansibirs, so we don't forget what we've lived and what we've learned. I know that for now this idea may sound like it's very far off, because thus it was introduced into our minds by the Egos in rebellion, but as we go about detaching ourselves from the Continuo (Continuum), we will understand that one lifetime is really an instant and that in the Trebolo we are prepared to enter into another matter which will have new experiences, with their joys and their sorrows."

"And how do you think people will react to the Konocimiento you express in your book, Helena?" Max inquired.

"I only need to transmit the Konocimiento as it was delivered to me and I transmit it as is; I haven't changed even one word, since these are the greatest truths that exist. Each person who reads it will be able to find one or many truths in it, depending on their intentions, because each one of us has the power to recognize the truth when we see it."

"Among so many, there'll have to be some brave ones to accompany us," noted Diego.

"Let's hope so, since this is the path on which every Ego must walk in order to know itself and its truths."

Helena turned to watch the sunset; the sun was hiding itself in the immensity of the firmament, projecting a multiple of tonalities that ranged from an intense red to a soft pink that was reflected in the clouds. She had never thought to receive such an enormous and complex responsibility. She was alone and she knew it very well. She was now the Maestra and she would strive to deliver the Konocimiento. She knew that in order to be able to do it, she needed the acceptance and cooperation on the part of humanity. This flow of energy between both parts was necessary in order to establish before the Konceptos the commitment to complete the instruction.

Helena thought about the dream she wanted to relate to them. Over the course of her life, she received many lessons and warnings in her dreams. Many of them were premonitions that were fulfilled. Today she had the hope that what she dreamed a few days ago would be able to be the future for everyone. The only thing that was missing was the desire to take the first steps in these transcendent moments on the Earth.

She said, "Now that we're leaving this town, we'll keep on walking, but in new reasons that we must bring closer to us. We've already clarified some of the causes of the past for which humanity hasn't been able to develop itself in the course of what should be. There are many more, but we must also project new ideas that vivify and nourish the Ego. I'm going to tell you about a dream I had.

At first, I felt I was flying in the air and I saw that the Earth was very different than it is now. New continents had been formed and humanity was divided into new races. I was able to see that there were cities on the Earth but there were also others that were suspended in the sky. Afterwards, I saw myself walking in a meadow of vibrant green, where the clarity and luminosity were extraordinary. There weren't as many people as now populating the planet. The people who lived there shared the environment in harmony with the animals that, for me, were completely different from those of today. The environment looked crystalline and the plants had an unusual height. I felt that the pure air that I breathed and the sound that came from the sun nourished and vivified my body, which was light and without the heaviness that we know. I moved from one place to another easily with my thoughts. I was admiring the surroundings when the scene changed and I found myself in one of those cities which floated in the atmosphere; the constructions were distinct from the ones I saw on the Earth. They were translucent, like prisms that filtered the light to then project it in ranges of colors and harmonious sounds. The people who lived in those floating cities collected the water from the atmosphere for their food. I understood they were in a different point of development from those below, and that the Egos were separated according to their levels of evolution, so they wouldn't hinder each other. I was seeing another HTime of development of the Planicio Terráqueo."

When she had finished recounting her dream, Helena felt a few drops of water fall from the sky, landing almost imperceptibly on her face. She burst out laughing, leaving Max and Diego amazed by the spontaneity and joy she radiated.

She recalled the voice of her Maestra Marion saying to her, "I'm going and I leave you all with all of your problems; we'll see at what point you understand. But don't worry; I'll always be with you all. When you feel warm water fall from the sky, you'll know it's me."

Joy is the act of living (27 Nov. 93).

Conclusion

When the idea of making this novel was born, it was with the purpose of delivering the Konocimiento (Knowledge).

We have to say that in the Prehistory and in the History of Humanity the Konocimiento has not manifested itself, therefore human perception was limited to what we call an appendix of Konocimiento, which is Sabiduria (Wisdom).

The Konocimiento is the Absolute, Sabiduría (Wisdom) the Relative. To our understanding they are the vital pieces; it is like an incomplete puzzle.

The pages of this novel explained what is specified exactly in this conclusion.

We must say that the Konocimiento is the Pleno Real (Complete Real) and Wisdom to a great extent is an Apparent, what we could say is a Fantasy (parodying).

Man has reached broad scopes within his meticulous acceptance of science, and digging here and there he wants to rummage around in the mysteries.

We do not live in an Absolute Reality but in a Projection of this that lacks consistency.

The Real within the Fusion grows until it forms a kind of Fantasy.

The Konsciousness of man is limited and subject to the Three-dimensional Reason.

Before entering the Planicio Terráqueo (Earth), which was ordered and conceived by our Padre Eterno (Eternal Father), to whom we owe respect and submission, the Egos who are we, were encapsulated in a Coarse Matter, but nevertheless marvelous, full of wonders. But it is not our destiny to continue in it since our condition of being in it was with the reason of comprehending and perceiving our past error.

We say that the Egos who arrived at this period was due our causes and our reasons, to us it is said Huestes of different levels of perception, ordered and subscribed before an Entity who was he who directed all of the Huestes; his error and his rebellion dragged all of the Huestes to a trial. This trial demanded the reparation and the creation of the Planicio Terráqueo so that it would be understood, this was defined in Seven HTiempos (HTimes) of Grace and birth within matter was in us. Previously and in Vital Principle we were manifested, now we are in the Reasons of Life which are projected in Birth, Growth, Reproduction, and Transformation.

We have the Trebolo which is when we leave our Matter to reappear again in another Matter and so on indefinitely until we understand.

The most revered Huestes became obsessed with having the command in order to complete three Erroneous Reasons:

1. Impose themselves and govern in order to drag the multitudes to their error and rebellious enterprise.

2. Lacking the capacity to comprehend since they do not accept or do not want to accept the returns in other matter, as they outline that they would lose their continuity, they seek immortality in the current matter, and for that end scientists work for salaries to "find" the great discoveries of Life.

3. In their arrogance and madness, lacking in their limitation, they seek to conquer the Universe or Fusion, without comprehending in their mediocrity that not everything is Real in the Fusion, but rather it is an Apparent. We use this word to define something that must be understood in its totality.

It is for this reason that we humbly wish to awaken to the Konocimiento in order to take a step forward towards complete understanding and not to the mistake to which our brothers have erroneously led us.

The Absolute